ANIMAL SPECIES AND THEIR EVOLUTION

HUTCHINSON'S UNIVERSITY LIBRARY
BIOLOGY

EDITOR:
PROFESSOR H. MUNRO FOX, F.R.S.
Professor of Zoology in the University of London

ANIMAL SPECIES AND THEIR EVOLUTION

by

A. J. CAIN

M.A., D.PHIL.

DEMONSTRATOR IN ANIMAL TAXONOMY,
UNIVERSITY OF OXFORD

HUTCHINSON'S UNIVERSITY LIBRARY
Hutchinson House, London, W.1
New York Toronto Melbourne Sydney Cape Town

First Published - 1954

Printed in Great Britain by
William Brendon & Son Ltd.
The Mayflower Press (late of Plymouth)
at Bushey Mill Lane
Watford, Herts.

CONTENTS

	Preface	*Page* ix
Chapter I	Introduction	11
II	Methods of classification	15
III	Rank	27
IV	Names	42
V	The polytypic species	48
VI	The biological species	73
VII	Other sorts of species	98
VIII	Geographical speciation	130
IX	Sympatric speciation	169
	Conclusion	182
	Suggestions for further reading	184
	Index	187

LIST OF ILLUSTRATIONS

Fig. I. Affinities of the gibbons, apes, and man 36

Fig. II. The evolutionary tree of the carnivores 39

Fig. III. Distribution and variation in the Great Tit 54

Fig. IV. Geographical variation in Rosellas 61

Fig. V. Interrelationships of different sorts of species 125

PREFACE

THIS book is intended to be a survey of the nature of species, their origin, and their evolutionary importance. The reader is assumed not to have any previous knowledge of the subject; consequently, some parts (notably the genetics of speciation) are presented in a simplified form, but I believe that no falsification has been produced thereby. More comprehensive and technical treatments will be found in the books and papers in the reading list (p. 184), especially those by Mayr and Dobzhansky who are the principal authorities in this field.

It is a pleasure to acknowledge the help and advice I have received from Professor H. Munro Fox, F.R.S., the editor of this series, from my friends Drs. J. H. Burnett, P. M. Sheppard and D. W. Snow, and from my wife.

A. J. CAIN.

Department of Zoology and Comparative Anatomy, Oxford

CHAPTER I

INTRODUCTION

THE science of biological classification is called taxonomy, and it is the task of the animal taxonomist to describe all the known forms of animals, to sort out their relationships so that they can be classified in as natural a way as possible, and to provide a system of nomenclature, so that each form can be referred to rapidly and accurately. In recent years it has been gradually realized that taxonomy is not merely a necessary pigeon-holing but also one of the most important activities in biology, requiring a synthesis of all other biological pursuits for its proper performance, and producing results of the highest importance in the study of evolution. That good classification is essential for all zoological work is obvious. No one would think much of a chemist who confused water and benzene "because they look alike". His classification would be defective, and might lead to serious practical consequences. Similarly, the equally incompetent zoologist who thought that whales were fish would soon find on further investigation that the apparent resemblance is misleading, and that in most features they agree with mammals. Or again, the forester who applies the remedies against a particular pest to his trees should be sure that it is that pest he is dealing with, and the physiologist investigating the 'common earthworm' will not confirm the results of others unless he is awake to the fact that there are several sorts of common earthworm, each with its peculiarities.

It is the business of the zoologist to study animals. But there are vast numbers of animals in the world, and they present an amazing diversity of structure, habits, and mode of life, as anyone can see if he will examine carefully a crayfish, starfish, jellyfish and dogfish. And if he will look at all the starfish, for example, that he can lay hands on, he will find there are many different kinds of starfish. Probably most people can tell apart several closely related sorts of bird or butterfly. But few know that there are forty sorts of earthworm in Britain alone, and probably several that, for lack of investigators, have never been described. Over half a million different forms of insect have already been described, and more are discovered almost daily. Even in Britain, one of the best-worked countries in the world, many kinds of animal remain to be discovered, and the fauna of huge areas in the tropics is hardly known at all.

Unfortunately, because of the difficulty of sorting this vast array of diverse forms, it has not infrequently happened that the taxonomist, studying a particular group and flooded out with demands for identification, has been unable to go further than naming his specimens and describing them in just sufficient detail to permit them to be distinguished from their nearest relatives. Often he has had to work entirely on preserved material which he did not collect himself and with very insufficient field-notes attached. And since a great deal of his time was taken up with finding the right name for a particular specimen, often changing well-known names in the process, there grew up almost a tradition among research workers in other fields of zoology that the taxonomist was purely a 'museum man' engaged in sorting skins or shrunken pickled specimens of animals he had never seen alive, using diagnostic characters that no one else could see, and wholly taken up with endless futile controversies about names. There was some little truth in this caricature. Even the mild and charitable Darwin said in a private letter, " . . . I have long thought that *too much* systematic work [and] description somehow blunts the faculties," although he added that the particular taxonomists he was writing about were "a very good set of men."

In the last twenty years there has been a gradual revolution. It is generally admitted by taxonomists on the one hand that to describe any particular sort of animal adequately one must take into consideration not merely its structure and distribution but also its genetics, mode of life, physiology, and all its other aspects. On the other hand geneticists, physiologists and students of evolution have realized the importance of good taxonomy, and in studying whole groups of animals at a time have been able to make generalizations of the utmost importance. In particular, the origin of species (a subject hardly touched upon by Darwin himself) is now far more clearly understood and is recognized as a most important stage in evolution. The old reproach that no one had ever seen a species evolving into others is no longer valid. And a comparison of the process of evolution as seen in action today with the fossil records of it in the past strongly suggests that the same sort of process acted then as now, and was as responsible for the main evolutionary lines as for the side-branches.

Until very recently, the species was thought of as merely one rank in a classification which was meant to bring together those animals that most resemble each other in small groups that could themselves be grouped into larger ones and so on upwards until one came to the group of all animals. In modern taxonomy several meanings of the word species can be distinguished, of which the most important is that it is the lowest group of animals which at least potentially form an interbreeding array of populations, unable to interbreed freely with other sorts of animals. It represents the attainment, in Dobzhansky's words, of "that stage of the evolutionary process at which the once actually or potentially interbreeding array of forms becomes segregated into two or more separate arrays which are physiologically incapable of interbreeding." The process of speciation is therefore of the utmost importance in evolution. Without speciation, specialization and the development of great efficiency in the exploitation of particular modes of living would hardly be possible, since the acquisition of special adaptations by some living organisms in one part of the earth would be continually hampered by outbreeding to others in adjacent regions. As climate, soils and waters vary all over the earth, the effects of selection to withstand particular conditions in one district would be nullified by constant interbreeding with organisms that had been subject to selection for withstanding the opposite conditions.

In this book, the older and simpler meaning of the species as merely one rank in the natural classification is first discussed (Chapter I-IV). Its extension, to take into account the extraordinary geographical variability of species (Chapter V) that was discovered when zoological exploration was begun on a large scale, was the beginning of the realization that structural differences are not always safe guides to species-limits, and that criteria of interbreeding in the wild are of greater importance. This realization has led up to the so-called biological definition of the species (Chapter VI.) This definition cannot be applied to all animals, and the treatment of asexually reproducing forms, and of 'species' that are portions of a single evolutionary lineage, requires special discussion (Chapter VII). Recognition of the importance of geographical variation, and the formulation of the biological definition of the species make it possible to show that probably the most important process in speciation is the production of geographically isolated populations which can become genetically so different while they are isolated that when they meet again they cannot interbreed freely. This is the theory of geographical speciation (Chapter VIII). And finally, mechanisms of speciation are considered (Chapter IX) that do not require geographical isolation, of which one certain example is known, and more may be discovered.

CHAPTER II

METHODS OF CLASSIFICATION

THERE are many different ways of classifying a set of things. We might arrange them according to whether or not they possess some particular character, and then subdivide the groups so formed by the presence or absence of another character, repeating this procedure until all the things in each of our smallest groups were for our purposes identical. This method would give a *hierarchy* of groups, each of which contained two others with mutually exclusive characteristics. For example, one could divide all living things into those that fly actively and those that do not. The group of fliers could then be divided into those with one pair of wings and those with two (since no animal with three is known). Those with one pair could be divided into those that have hair (the bats) and those that have no true hair (birds, many insects), and so on. The virtue of such a classification as this is that if the characters used for diagnosis are selected for being conspicuous, it will be extremely easy to work through the classification and find out to what group any particular specimen belongs. Such classifications are therefore used for identifying and are usually called 'keys'. In the example just given every group contains two others. It is a *dichotomous* key. Sometimes a group may be divided into three or more according to convenience. One could divide the group of fliers with one pair of wings into those with hair (bats), those with feathers (birds), and those with jointed bodies (insects). But in general, the dichotomous key is used, as being the simplest and easiest to work through.

Alternatively, one could remove from one's set of objects all those with some striking peculiarity into one group, those with another into another and so on, until one had a large number of mutually exclusive groups of equal status, instead of a hierarchy. The difficulty in this procedure is that one may have to read through descriptions of all the groups to identify one specimen, and that the last group is usually a rag-bag full of all those objects without obvious distinguishing characters. In the insects, the group of *Neuroptera*, until it was restricted to the lacewing flies and their allies was just such an assortment, and the old classification, so well known to everyone, of the *Lepidoptera* into butterflies and moths distinguished one fairly clear-cut group (the butterflies) while leaving together a large array of others (the moths) each of which is now correctly accorded at least equal status with butterflies. Another difficulty is that, particularly in sets containing large numbers of closely related forms, it is hard to find a series of diagnostic characters of which all specimens have one and one only. For example, one might group the British butterflies into those having some white somewhere on the wings, those with some yellow, those with some blue, etc., but some have both white and yellow, or both yellow and blue. The difficulty is intensified when the males and females are remarkably different. One way is to prescribe the order in which the characters are used, and that transforms this method into a hierarchical method straight away. For example, one could divide butterflies first into those with some and those with no white, and then each of those groups into those with some and those with no yellow. Often, the taxonomist who has studied a particular group well can say on looking through a key that it was clearly constructed by removing all the obvious subgroups first, and was then licked into a dichotomous shape. The symptom is the rag-bag group at the end. Rag-bag groups always give trouble except in keys, which are made solely for convenience in identifying. They are defined solely by what they are not, not by what they are, and nearly always they turn out to be very miscellaneous groups.

The two methods just given produce mutually exclusive groups, whence their utility for keys. Either a specimen has the distinguishing character of a certain group or not, and if not, it must go into some other. But they rely at each division on the use of a single character. Unfortunately, otherwise unrelated animals may have characters in common. Both bats and butterflies have wings, but bats are vertebrates, butterflies are insects. They differ fundamentally in all other characters. And with objects as complex as animals it very frequently happens that in one or two members of a group which is otherwise very clearly defined, that very character which one would like to use to diagnose the group is absent, although all the other characters of the organism concerned leave no doubt that it is a member of the group. For example, the group of birds (warm-blooded vertebrates, with feathers, and laying large yolky eggs) is a very natural one, and the possession of a pair of wings for flight is one of its most characteristic features. One can recognize a flying bird as a mere silhouette a mile off, when its possession of a backbone, feathers, warm blood and the capacity for laying large yolky eggs are not readily discernible. But the kiwi, for example, cannot fly. Yet it is certainly a bird. Here, one can remove the difficulty by simply changing the diagnostic character, and using not flight but feathers. Again, everyone is accustomed to the idea that mammals are warm-blooded vertebrates with true hair (like we have ourselves), that bring forth their young alive, and suckle them by means of mammary glands functionally developed only in the female. In the echidna or spiny ant-eater, and the duck-billed platypus, the hair and mammae are present, but an egg is laid. In this respect and in many features of the skeleton, they are closer to some extinct reptiles. However, on the whole, their features are mammalian, and they are accordingly grouped with the hedgehog, the bat, the whale, the cow, the cat and ourselves. The difficulty here is rather more acute, as their inclusion makes the mammals a more heterogeneous, less easily defined group. In some series of fossils it becomes intolerable. In many series of fossil mammals, although the latest forms are very distinct and specialized, more generalized forms are found in older rocks, and finally one gets down to a mass of very closely related forms. In each series leading from this mass to the end-forms, there may be so profound a change that no single character remains unaltered and available as a diagnostic, except those that are too broad for use within the group. For example, in the great group of primates (tree-shrews, lemurs, monkeys, apes and man) it is almost impossible to find a single diagnostic character (running through the whole group) which is not so basic that it says only that all are mammals, and therefore fails to divide them from hedgehogs and other insectivores.

In short, reliance on a single character will not only group together unrelated forms (as when in our first example we lumped bats, birds and flying insects merely because they all fly) but may even get us into a position where we can produce no diagnoses at all, and have no one character to rely on, although the group we are dealing with is obviously a good one—that is, its members have more in common one with another than they do with members of other groups. It is essential, if we wish to group similar animals together, to take account of all features, and look for general resemblances rather than particular differences.

This changeover from classification by difference to classification by resemblance is of the utmost importance. The distinction between the two methods was clearly understood by some earlier naturalists, especially the great taxonomists John Ray (1627–1705) and Linnaeus (1707–1778). Indeed Linnaeus, while establishing a classification of plants employing effectively diagnoses by single characters, and providing keys so good that they are still used in schools in Sweden, explicitly sets out the idea that in the most *natural* classification (not the one most easy to use for identification) all possible characters must be taken into account, no single one being used intolerantly. This 'best' system may be referred to as the *natural* system of classification, all others being called *artificial.* In the natural system, animals are grouped according to their basic similarities into as many groups and subgroups as their resemblances and differences require.

It is most important to be quite clear about the meaning of the adjectives 'natural' and 'artificial' as applied to biological classifications. They mean no more than is explained immediately above. Every possible intermediate between an obviously artificial and an obviously natural classification can exist. All depends on the number of characters taken into account. And it should also be noted that the construction of a natural classification does not necessarily imply any theory about the nature of the relationships between the groups included in it. Birds, for example, can be recognized as a natural group closely related to reptiles and mammals, without thereby postulating a common ancestor, or any particular theory of descent. This is the reason why Darwin's friend, the great botanist Hooker, was able to accept the theory of evolution without making any change in his taxonomic labours. He had merely to postulate descent from a common ancestor instead of 'affinity' or 'archetypal resemblance' to explain resemblances between natural groups.

But although all characters must be taken into consideration, they are not treated as of equal importance. The general shape of a whale is regarded as a necessary consequence of its mode of life, and therefore of no special importance, whereas its possession of warm blood, the mammalian type of heart and arrangement of main arteries, lungs, and mammalian methods of reproduction, is held to show unequivocally that it is a mammal, more closely allied to the cow, the lion, and man than to any fish. Its basic theme, so to speak, is mammalian, and it shows the particular variations necessary for a permanently aquatic life such as fish live.

When an animal is profoundly modified for a particular mode of life, very few characters may be left to indicate its general plan, and they may seem at first sight to be quite trivial. The horse, donkey and zebras form a very natural group of mammals, grazers or browsers with long legs for running, hoofs, complex grinding cheek-teeth, and other features in which they resemble the very natural group of cattle, sheep, goats, antelopes, giraffes and deer. At first sight the foot of a horse and of a cow are much more alike than is either to the paw of a lion, for example. Yet a detailed examination of horses and cows and especially of their fossil relatives shows clearly that the features in which they resemble each other are adaptations to a very similar mode of life. The horse, the tapir and the rhinoceros are far more closely related than the horse and cow. The mere possession of hoofs is not a diagnostic character of a natural group of mammals. Nor is the development of a thick skin. For a long time the horses, asses and zebras were classified with cattle, deer and antelope in the group of hoofed mammals or Ungulata, and the rhino, hippopotamus and elephant were sometimes put together in a single group of pachyderms (thick-skins). But the elephants are the last survivors of a quite distinct and formerly enormous group of mammals, the hippo is closely related to the pigs and thence to cattle, and the rhino and tapir should be associated with the horse. A group of horned mammals including the rhino and cattle would be as artificial as the old group of hoofed mammals. But the group of 'cloven-hoofed'—that is, two-hoofed—mammals is a natural one.

To see how natural groups are delimited, let us consider the group of Carnivores, which includes such forms as the polecat, weasel, fox, badger and wild cat in Britain, the raccoon and skunk of North America, the bears, wolves, great cats, mongoose, civet, giant panda, kinkajou, and a host of other forms. If we take the group as a whole there is obviously great diversity within it, but certain features stand out. Almost all are carnivorous in diet, and hunt their prey. Consequently almost all have the means for rapid locomotion, for holding the prey, and for tearing it to pieces. The front teeth are therefore well developed, both for cutting (the incisors) and for stabbing and holding on (the canines). The teeth further back, even those that are almost always developed for crushing in other groups (the molars) are high and sharp, and the hingeing of the lower jaw with the skull is so arranged that there is very little sideways movement possible. The jaws tend to function, in fact, as flesh-cutting shears. As the limbs are of value in holding down the prey, they are nearly always capable of complex movements, and the digits (which are armed with claws) are never reduced to less than four. A useful diagnostic feature is that three of the wrist-bones (the scaphoid, lunar and centrale) are always fused together—exactly why is not known, but so little work has been done on the mechanics of the wrist and ankle in any animals that this is not surprising. Also, the collar-bone or clavicle is reduced or absent, but this condition is met independently in many other groups, and although characteristic is not diagnostic.

Now, exceptions can be found in living forms to all the main characters of the group (except the fusion of the wristbones). The hind-teeth, except the main shearers are often reduced in number. In the dog, for example, almost the full set is found, there being in each half of the upper jaw three incisors, one canine, four premolars, and three molars (which are not preceded by milk-teeth as are all the others). In the lower jaw there is the same arrangement except that the last molar is absent on each side (rarely present as an individual aberration). In the lion, on the other hand, the incisors and canines are unaltered but there are only three premolars on each side in the upper jaw, two in the lower, and only one molar in both upper and lower jaw. Of these, the last upper premolar and the first (and only) lower molar are greatly developed as shearing teeth or 'carnassials', as they are in the dog although not to the same extent.

In the aardwolf, *Proteles*, a small slim hyena-like African carnivore, related to the civets, genets and mongoose, the jaws are feeble and the teeth small and very uniform in character. But this animal is an insect-eater, so the difference is not surprising. Again, the skunks, weasels, otters and badgers are closely related, and reduction of the back teeth has taken place here as also in the cats, in association with a flesh-eating diet. In the badger the molars are greatly reduced in number, but the shearing teeth have become considerably broader, for crushing, and the diet is very generalized. This change appears to be due to a secondary adaptation to being omnivorous. In the large-eared fox of South Africa (*Otocyon*) the teeth have actually increased in number, so that there may be five molars on each side below and four above. Why this increase beyond the usual maximum has taken place is unknown.

Shape and size of the body do not seem to be particularly characteristic of the *Carnivora* as such. Several heavy flat-footed tailless types are known; the bears are one obvious example. Another is the giant panda which is very bear-like, but not at all closely related to the true bears. Its kinship is with the raccoons of North America and the kinkajou. It is further remarkable in feeding almost entirely on bamboo shoots. Its molar teeth are correspondingly rather broad and therefore better adapted for crushing. The kinkajou, a most delightful animal, is also remarkable, for a carnivore, in having a prehensile tail, which makes it the most perfectly arboreal of all carnivores.

Again, the dog and the wolf are built for running down their prey; in consequence, their legs are long and the feet are (in proportion) small—very different in general build from the legs and paws of the cats, which creep up under cover and spring on their prey. But one undoubted cat, the cheetah, is also a swift running hunter, and its legs and feet are so remarkably dog-like that within the group of cats it is usually kept in a subgroup to itself.

In short, the characteristics of different carnivores are correlated with their modes of life, and many of their most obvious features are adaptations. Only the teeth and limbs have been particularly mentioned, but corresponding variations could be found in the digestive tract and musculature, for example. Now, when a character is seen to be clearly adapted to a particular mode of life, its importance in the classification of a group is greatly reduced, unless it is also diagnostic of the group.

For example, the seals, sea-lions and walruses show clearly by the fusion of the scaphoid, lunar, and centrale in the hand that they are closely allied to the rest of the carnivores. But the tail is reduced to a mere stump, the limbs are converted into flippers, and the teeth are comparatively simple (as is usually the case in fish and mollusc eaters). They show none of the special features of other groups of mammals, and seem to be carnivores highly modified for an aquatic life, although not so modified as are the whales or were that remarkable group of marine reptiles, the Ichthyosaurs. The possession of flippers happens to be diagnostic within the carnivores for this group, and they are used for propulsion, not merely as steering-organs as in the porpoise or other whales. Now, the otter, which from the details of its skull is closely allied to the stoats, skunks, and badger, and is more like them in general proportions than like a seal, is also aquatic, with webbed feet, with which it swims, and can close its nostrils and ears under water. It is approaching the seal-type, and one sort is actually marine. A little more adaptation to a more purely aquatic life, and there might well be a second stock of seal-like carnivores in the world, in which case the important features of the true seals would be not flippers but some inconspicuous group of characters, not obviously correlated with their mode of life but indicating their affinities with related groups. If either the mode of life causes a modification of every feature of the animal, or the two stocks that show parallel adaptation were very closely related in the beginning, it may even become impossible to decide which features, if any, can be used as clues to their true affinity. It is for this reason that natural classifications often employ obscure features of apparently no importance whatsoever, and that keys may bear little relation to natural classifications.

Then what are the characters used in classifying the modern carnivores? Because of this multiplicity of forms adapted for different modes of life—a common phenomenon in all groups of animals, referred to as *adaptive radiation*—there are very few. The group is characterized as being a group of true mammals primarily specialized for flesh-eating, and normally having well-developed front teeth and shearing back teeth, in which the shears are formed from the last upper premolar and the first lower molar, and in which three of the wrist-bones are always fused; plus a group of aquatic forms with reduced dentition and flippers, which have no teeth modified as shearers, but do show the fusion of the wrist-bones.

In fact, the carnivores are characterized only by a complex of features of which many may be absent in particular examples. Now, if we look at the fossil forms, we find that about fifty million years ago there existed a small group of animals, the miacids (rather badly known because they seem to have been tree-livers, which are unlikely to become fossilized readily), which resembled small modern carnivores quite closely; in particular there were carnassial teeth, and they were the same teeth as in modern forms. But the wrist-bones were not fused, and there were other features in which these forms were less specialized. Side by side with these were another and much more abundant group, which seems to have died out when the modern forms and their extinct allies became common. These, the creodonts, were even more generalized. The complete set of teeth was almost always found, and when carnassials were developed, as they often were, they were teeth lying further back than those used in the miacids and modern forms. The brain was smaller, the teeth more generalized in structure; both pairs of limbs had the full complement of five digits nearly always.

In the creodonts there was considerable adaptive radiation, since a few became large as bears and remarkably bear-like in their teeth, some were cat-like, others weasel-like, and one jaw-bone strongly suggests a form similar to the sabre-toothed tigers. But the earliest and least specialized can hardly be separated on any character from that very generalized and basic group of mammals, the *Insectivora*, which survives today in the shrew, the mole, the hedgehog and other related forms, and which approaches also the least specialized primates, and the simpler forms of many other groups of mammals. So when we take the fossil forms into consideration we find a good approximation to a chain of forms leading back from undoubted carnivores to forms which are too like the group called the *Insectivora* to be placed certainly with either. There is no one diagnostic feature which will separate out all the forms in this chain from all other forms. The middle members resemble the more specialized in some features and the less specialized in other, more general, ones. No feature is constant throughout, but there is a sufficient overlap of features for any one form to be put in its place, so that the chain can be built up. The importance of any character in the group is assessed by testing its constancy within subgroups constructed by considering all other characters.

Thus all the skeletal characteristics of the miacids (all one has to work with in fossils) agree well with those of the least specialized modern carnivores, i.e. with those that are neither huge stalking hunters like the lion, nor runners like the dog, nor insectivorous like the aardwolf, nor aquatic forms like the seals, except that all the bones of the wrist are separate, and there is no auditory bulla, that is, the middle ear is not protected by a special capsule of bone. Since all other characters agree, and since these two characters indicate only that the miacids are less specialized because they share them with the definitely less specialized creodonts, it seems that the miacids and the modern carnivores are very closely related. In fact it is difficult to draw the line between them. The real use in taxonomy of the fused wrist-bones, whatever their functional significance, is as a diagnosis among living or recent mammals. That an identical fusion is found in the seals, with their very different mode of locomotion suggests not only that there is a real similarity between seals and other modern carnivores (a suggestion not controverted by any other feature) but also that this particular character, being unaffected by the profound difference in locomotion may be expected to be a very stable one. The first suggestion seems quite correct, but the second is less happy; it is likely that any animal with this particular fusion would certainly be a carnivore, but its absence need not prove a particular animal to be something else. As remarked already, the maximum number of molars in the groups of true mammals is almost invariably three, but *Otocyon*, an obvious carnivore, has more. Intolerant use of this particular character might cause *Otocyon* to be separated from all other mammals. No one who has seen the beast (it is a most attractive fox-like creature with huge ears) could possibly take this step, yet less than a hundred years ago a most competent anatomist proposed to separate man from all other true mammals in a group by himself, on evidence that was very little better.

The fusion of these three wrist-bones, then, is probably

unimportant as a separating character (although of great importance in tracing resemblances) and in view of the other resemblances need not separate the miacids from the modern carnivores, since it is quite possible that it may have arisen several times in closely related stocks. But it is highly important in indicating the relationships of the seals. Similarly there is one extraordinary lemur in Madagascar, the Aye-Aye (*Daubentonia*) that has huge gnawing front teeth exactly like those of a rat, mouse, or squirrel; in fact, when first discovered, it was described as a sort of squirrel. But in every other respect it belongs with the lemurs, a rather unspecialized group allied to the monkeys and included with them, the great apes, and man in the group *Primates*. The possession of rodent incisors is therefore of very little importance in the primates, but is a constant and highly important feature in the huge group *Rodentia*, which includes squirrels, rats, mice, voles, porcupines, agoutis, capybaras, guinea pigs and all similar gnawing animals.

From the examples just given several conclusions can be drawn:

(i) Natural groups are recognized by the general resemblance of their members, irrespective of variation in any single character.

(ii) But great resemblance may be produced by independent adaptation to the same mode of life. All characters attributable to such convergence must be disregarded in determining the relationships of their possessors.

(iii) In some natural groups all the members of some subgroups may be found to possess a certain character in common; this character will then have considerable taxonomic value. In others the same character may be very sporadic in occurrence, and of very little taxonomic value.

(iv) One natural group may be easily recognized by a peculiar character or complex of characters found in all its members and nowhere else. Such a character or complex is then diagnostic. Another group may be equally natural yet have no single diagnostic feature.

In fact, the taxonomic importance of any character within a group depends entirely on how far its occurrence within that group is correlated with all other characters.

It has often been suggested that the best taxonomic

characters are non-adaptive ones, since these are least likely to vary because of the different modes of life of different members of a group, and will remain unchanged as evidence of their mutual affinity. This is only half true. Certainly those characters most easily affected by a slight change in mode of life are not likely to be useful as pointers to their owners' general resemblances. But it does not follow that those characters which remain constant throughout a group, since they are not affected by changes in mode of life, are not adaptive. The gnawing apparatus of the rodents, for example, shows only small variations throughout this huge group, the most successful group of living mammals, yet it is clearly an adaptation to handling hard as well as soft food. Nor is it true to say that very trivial characters cannot be adaptive. Many 'trivial' characters appear so to us simply because of our ignorance, and others are merely the most obvious products of genetic complexes whose other effects are of great importance to the organism.

To sum up, the many different sorts of animals can be classified in many ways for many different purposes. When all possible characters are taken into consideration, they are said to be classified on the natural system, and this, reasonably enough since it has the widest basis, is the system used in animal (and plant) taxonomy. Any one natural group may contain any number of subgroups according to the degrees of mutual resemblance of its constituents. Classifications designed for rapid identification and based on the most convenient characters are called keys and are usually dichotomous. They are one sort of artificial classification. It is not always possible to arrange natural classifications as keys since some natural groups may not have a single diagnostic character. Taxonomic characters are merely those which after consideration of all the characters of a group are found to be most useful in writing down a definition of it.

CHAPTER III

RANK

If for the moment we confine ourselves to living forms, then it is true to say that classification by the natural system gives us a number of groups, which themselves can be collected into larger groups, and so on upwards until we come to the group of all animals, and downwards until we come to individual animals. For example, man himself is so obviously a mammal, and so close to the great apes and gibbons, in all but a very few respects (admittedly most important ones to him) that zoologists, who know that there is more difference in every way between a jelly-fish, a starfish and an eel than there is between an eel and a man, often have difficulty in understanding the attitude of some non-zoologists on this point. In a recent revision of the mammals by Dr. G. G. Simpson, man, with certain fossil forms, is one sub-group of a large group that contains the gibbons and great apes as a second subgroup, the old-world monkeys as a third, the marmosets as a fourth, and the rest of the new-world monkeys as a fifth. Of these, the marmosets and the other new-world monkeys are closely related, as are the gibbons, great apes, and man, so three subsidiary groupings are recognized which contain first the marmosets and new-world monkeys, second, only the old-world monkeys, and third, the great apes, gibbons and man.

Now it is obvious to anyone that a group which contains only the chimpanzee and orang-outang is a much smaller and more homogeneous group than one containing both these and dogs, bears, skunks, weasels, otters, civets, camels, cows, whales and all their relatives, and must be ranked low in the hierarchy of the natural classification. And when a number of groups are being discussed it is very convenient to give some idea of how far up in the hierarchical classification they are, because this gives an indication of how homogeneous they are. The naming of particular groups, natural or unnatural, is of course of extreme antiquity. Aristotle knew the group of whales as *Cete*. He correctly separated it from the fishes and placed it near the mammals—a procedure not followed for nearly two thousand years. Other names of particular groups are as old. But Aristotle appears to be the first to employ *rank-names* to designate all natural groups of about the same status, and to discuss criteria for deciding the rank of a particular group. Linnaeus, however, was the first to provide a really comprehensive scheme, and the rank-names used in biological taxonomy derive mainly from him.

Linnaeus's hierarchy began with the whole universe as known to us, which he called an *empire*. This rank has fallen into disuse. The empire was divided into three *kingdoms* (animal, vegetable and mineral). Each kingdom was divided into *classes*. In animals, there were six of these: Mammals, Birds, Reptiles, Fish, Insect, and Vermes ('worms'), the last two now very extensively revised and broken up. The next rank down was called the *order*. For example he distinguished eight orders within the class of mammals, of which only three are now recognized as natural. These were the *Carnivora* (including the seals) which he called the *Ferae*, the cloven-hoofed animals or *Pecora*, now called *Artiodactyla*, and the whales or *Cete*, now called *Cetacea*. All the others have been broken up, and several mammalian orders are now known from fossil material discovered long after Linnaeus's time.

Orders were subdivided again into *genera* (singular, *genus*), these into *species*. The species consisted of the most homogeneous groups it was possible to make, each clearly separated from the rest. But when some individuals of a species, while clearly belonging to it, showed some marked peculiarity, they were recognized as a distinct *variety*. Linnaeus's hierarchy of ranks, then, was *empire*, *kingdom*, *class*, *order*, *genus*, *species*, *variety*. With the great increase in number of the sorts of animals known to zoologists, it was inevitable that additional ranks should be employed, and that most of Linnaeus's groups should be revised and subdivided. In particular, his genera were far wider in range than most of those recognized today.

The break-up of Linnaeus's rag-bag groups of 'insects' and 'worms' showed that within these classes there was more diversity than in the other four (mammals, birds, reptiles, fish) taken together, and new ranks between the kingdom and the class were introduced. The breaking up of many genera also necessitated new ranks between the order and species. At the present, a large number of ranks may be used but every specimen must be referred to a particular *species*, this to a particular *genus*, *family*, *order*, *class* and *phylum* (plural, *phyla*) within the animal *kingdom*. These are the obligate ranks or categories. It is open to any taxonomist who feels he has reason to recognize any number of ranks within this scheme that he needs to express his idea of the classification of a particular group. For example, in the class *Mammalia*, the order *Primates* contains, in Simpson's classification, 158 genera, mostly extinct, grouped in eighteen families. These fall pretty clearly into two main groups (some people would prefer to split up one of them at this level and make three). These are referred to as *suborders*. The suborder *Anthropoidea* contains the monkeys, apes, and man. This is the group referred to at the beginning of this chapter. In it, man and his fossil allies form the family *Hominidae*, the gibbons, chimpanzee, orang-outang and gorilla together with seventeen extinct genera constitute the family *Pongidae*, and an extinct genus of apes, the very unspecialized *Parapithecus*, is the only genus in a third, the *Parapithecidae*. All these, which are closely related, are grouped in a *superfamily*, namely, the *Hominoidea*. The other two superfamilies in the suborder are the *Cercopithecoidea* with only one family (the old-world monkeys), and the *Ceboidea* with two families, the *Callithricidae* containing the marmosets, and the *Cebidae* containing all the rest of the new-world monkeys. When, as often happens, the genera within a family fall into two or more groups, then *subfamilies* can be recognized, and so on. All living men are classified in a single species, which with several extinct ones make up the genus *Homo* (Latin for man). This, with other genera is in the family *Hominidae*, and the complete classification of man from the genus upwards is as follows:

Genus	*Homo*	
Family	*Hominidae*	
Superfamily	*Hominoidea*	
Suborder	*Anthropoidea*	
Order	*Primates*	
Infraclass	*Eutheria*	(true mammals, as contrasted with marsupials and other forms).

Subclass	*Theria*	(as contrasted with certain very remarkable extinct forms close to some extinct reptiles).
Class	*Mammalia*	(mammals)
Subphylum	*Gnathostomata*	(animals with backbones and with true jaws, including mammals, birds, reptiles, amphibia, and most fish).
Phylum	*Chordata*	(all animals with backbones, and their allies).
Subkingdom	*Metazoa*	(all animals except the sponges and the uni-cellular animals which form separate subkingdoms).
Kingdom	*Animalia*	(all animals).

This classification brings Man progressively into relation with all other animals. Its convenience is extremely great. When one has learnt the general characters of the major groups of animals (phyla and classes), then if a particular genus is mentioned one has only to ask "In which class?" and the answer allows a great deal of information to be attached to it. For example the sentence '*Pulex* is secondarily wingless' tells us little. But if we add (*Insecta*) after *Pulex*, then *Pulex* is a true insect. That is, it is a complex animal with a hard secreted external skeleton, normally six jointed walking legs, compound eyes, with a main nervous system running down the underside of the body (not down the upper as in ourselves and all other vertebrates), with a unique system of tubes carrying air direct to the tissues of the body, with an elongate simple dorsal heart, and many other special features. In fact, *Pulex* is the common flea, which is wingless but undergoes a metamorphosis from egg to larva to pupa to adult exactly like that of the butterfly, beetle or bee.

The principal rank-names in use today are listed below in hierarchical order. Only the obligates (in italics) must be used in classifying any particular group. There might even be a single species so different from all other animals that it occupied a separate genus, this a separate family, and so on right up to a separate phylum. On the other hand, in the insects some families have hundreds of genera and thousands of species, so that many ranks may be employed even within a genus.

Kingdom, Subkingdom, Grade, *Phylum*, Subphylum, Superclass, *Class*, Subclass, Infraclass, Superorder, *Order*, Suborder, Infraorder, Superfamily, *Family*, Subfamily, Tribe, *Genus*, Subgenus, *Species*, Subspecies, Variety.

As all these are ranks of natural groups, the term 'group' can be used to refer indiscriminately to examples of any of them, and to any groupings which it is not desired to name specially. For example, the term 'species-group' is often used for a number of species within a genus which are closely allied. In addition, there are certain rank-names which have been applied by different authors in different ways. Such are *brigade*, *cohort*, *legion*, and *section*. However, no biologist uses these without a clear indication of where he inserts them in the customary hierarchy, and there is rarely any difficulty. I wish the same could be said of some scientific popularizers, who will cheerfully use variety and class, for example, as synonyms, and some physiologists who speak about differences between species, when they mean differences between a species in one order or class and another in a different order altogether.

The system of nomenclature is discussed below, but it should be noted here that the names of genera are treated as singular. "*Pulex* is a genus of flea." Those of all other groups are plural, so that one can say "the Class Mammalia *is* closely related to the Reptilia" or "The Mammalia *are* closely related . . ." but not "The Mammalia *is* related . . ." Consequently, it is easy to see whether a particular name with no rank attached is of a genus or not. Also, by convention, the names of some groups have standardized endings. These are as follows:

Superfamilies	-oidea
Families	-idae
Subfamilies	-inae
Tribes	-ini
Subtribes	-ina

Examples given above are, *Hominidae*, *Pongidae* (families), *Hominoidea*, *Ceboidea* (superfamilies). Groups in ranks above these have no prescribed endings, and occasionally a few may be found with the ending prescribed for superfamilies—for example, the suborder Anthropoidea, mentioned above. But these rarely cause any confusion.

This is the system of ranks in universal use in zoology. Let us now consider just what a rank is. How does one recognize not particular orders or families, but orders as such or families as such? What are the characteristics of the Order or the Family? This question is best answered by considering exactly what one does in practice when classifying.

The Anthropoidea, as already mentioned, are a group of mammals containing the monkeys, gibbons, great apes, and man. These form a natural group in the strict sense. All are obvious mammals. All have large eyes which face forwards and give a wider field of stereoscopic vision than in any other order of mammals. Also, there is a special part of the retina which is adapted for seeing detail clearly. In all, the orbits (eye-sockets) are protected by a complete ring of bone. In comparison with almost all other mammals, the snout (and with it the sense of smell) is greatly reduced. The limbs are very flexible, the digits capable of complex movement and rarely reduced. In most, both the thumbs and great toes are opposable, so that either the feet or the hands or more often both function like the hand of Man. There is no tendency to fusion of any of the limb-bones. The collar bone is always present and well-developed. In almost all there are flat nails, not claws. Almost all are arboreal, and the few non-arboreal members show many signs of affinity with arboreal forms. The dentition is reduced, there being only two pairs of incisors in each jaw, and three or two pairs of premolars. The teeth are not highly specialized as they are in the rat, horse, lion, or elephant, and the diet tends to be omnivorous. One of the most characteristic features is the large rounded braincase, containing a brain which, relative to the size of the animal, is strikingly large even at its very smallest. Because of this expansion the face is only a small part of the head, and the foramen magnum, or opening in the skull through which the spinal cord joins the brain, tends to look downwards rather than directly backwards as in all other mammals.

Within this natural group, the rank of which does not concern us at the moment, other natural groups can be discerned. If we take all the known specimens—which are always the basic evidence that any taxonomist must work with—they can be sorted into two fairly obvious groups which happen to be geographical groups as well. As a useful diagnostic character (not always very well developed) the South American monkeys have nostrils separated by a thick fleshy partition and tending to look outwards. In the old-world monkeys, apes and man, the partition is a thin one so that the nostrils are close together, and they open downwards, not outwards or sideways. For this reason, the first are usually referred to as the Platyrrhines (flat noses) and the second as the Catarrhines (downward noses). The Platyrrhines are on the whole rather small. Almost always there is a long tail, which may even be prehensile. All are thoroughly arboreal, and in many the hand is long, a hook for swinging rather than grasping, and the thumb is not easily opposable to the other fingers. The dentition, except in the marmosets, is less reduced than in old-world monkeys, since one more premolar is found (on each side of each jaw). In the marmosets, the last molar is lost, so that the number of teeth is as in the old-world monkeys, although the arrangement is as in the other new-world forms. The marmosets are further distinguished by their small size, squirrel-like proportions, and the possession of claws on all digits except the big toes. These features might be regarded as unspecialized, but the teeth are not. It is more probable that the marmosets are secondarily specialized for a squirrel-like mode of climbing (digging the claws into branches and boles too large to be grasped) than that they are truly unspecialized. All known fossils of the Platyrrhine monkeys are also South American. The group seems a natural one, within which the marmosets clearly stand out from the rest.

The Catarrhines are entirely old-world. These tend to be larger forms, not so well furred, and often with hard fleshy pads for sitting on. The premolars are always reduced to two pairs (i.e. eight, there being two on each side of each jaw). Several forms, including man and the baboons, are more or less terrestrial. The tail is often short or even absent externally. The big toe is always well developed, and opposable except in man; the thumb is usually opposable. Many forms have remarkable cheek-pouches for storing food. In all, the external auditory meatus (the passage leading from outside to the eardrum) is bony.

The anthropoid apes and man resemble one another in the

structure of the brain, the tendency to develop an upright posture in walking, the absence of a tail and in many important details of their physiology. It is generally agreed among students of the group that the difference between man, the great apes and gibbons on the one hand, and the old-world monkeys on the other, is greater than that between the marmosets and the rest of the platyrrhines. Within the great-ape group, there are three obvious subgroups. There are several sorts of gibbon, all highly arboreal, with extremely long arms by means of which they swing with remarkable speed from one branch to another—a mode of progression called brachiation. The hands are very long and used as swinging-hooks but the thumbs are well developed. They can walk upright on large branches or the ground, holding out their arms as balancers. Sitting-pads are present.

In contrast to the gibbons are the gorilla, chimpanzee and orang-outang, all much larger, with long but not inordinately long arms. They are primarily arboreal, but the gorilla spends much time on the ground, walking with the support of the arms. The brain, large in the gibbons, is even larger in the apes. There is still something of an elongate muzzle, with the teeth behind the canines arranged in two parallel rows, almost as in a dog. In the chimpanzee and gorilla, remarkably enough, one of the wrist-bones, the centrale, is absent, as it is in man, although present in all other primates.

In man, the brain is huge and contains some special centres not found in the apes. The teeth are arranged around a continuous curve, not in two parallel rows, and the legs and feet are clearly specialized for bipedal progression. The rather fragmentary fossil evidence indicates clearly that man, the great apes, and the gibbons are the present terminations of three distinct lines of specialization, which appear to converge on each other as older and older fossils are considered.

Now, suppose we are confronted with a particular individual 'monkey' and told to classify it. It is obviously an animal. It has a nervous system, muscular system, and digestive tract, and eats complex foods. It is clearly not a plant. At the top end of the hierarchy, then, we recognize the group of animals by a certain number of peculiar characters. We can instantly put our specimen into the Kingdom *Animalia*, and must note that what we are doing is classifying it by those special characters which distinguish it from plants or minerals with all other animals. The rank-name for this group of animals could be 'Republic' without affecting the grouping in the slightest.

At the bottom end of the hierarchy, it is an individual, and we cannot divide it further taxonomically. The idea of individuals is thoroughly familiar to us because nearly all the living things we know well are independent to a very large extent of rigid spatial relations with other living things. In some of the less generally familiar groups of animals, however, there is every possible gradation between undoubted individuals, individuals that commonly remain close to one another, individuals that are physically connected with one another, often in elaborate patterns, such as corals, and colonies in which the connexion is so intimate and so much of the functions of life are carried on by the colony as a whole, that the colony is now the individual. However, if we adopt the criterion of permanent rigid spatial dependence for the parts of an individual we shall not go far wrong. The individuality of a particular specimen, then, is recognized by comparing it with other specimens and determining whether it is divisible further without loss of individuality or not. A single monkey is the smallest unit we can possibly recognize in the group of all monkeys as being a complete individual; a ring of monkeys holding hands is not.

Now, as soon as we start grouping individuals according to the natural system, we do so according to their peculiar characteristics as compared with those of all other animals. Suppose that the 'monkey' in question has no tail, distinct opposable thumbs and big toes, extremely long arms, only thirty-two teeth, long silky black hair, and the second and third toes jointed together by a peculiar web of skin almost to their free ends. These characters would suffice immediately to characterize it as a male siamang, the largest form of gibbon. It differs from the other sorts of gibbon both in the webbing of the toes, and in possessing a large dilation in the throat which gives resonance to its howls. Consequently, it is reasonably separated from the other gibbons in a group by itself. The gibbons, as already mentioned, form a natural group allied to but differing from the great apes and man, and so on. If we make a diagram in which points stand for actual individuals that have been examined, and their distance apart is some measure of their affinity on the natural system then for the gibbons and their allies we shall get something like figure I. In this figure, the clustering of the dots around particular places shows that all the known specimens of these animals tend to group clearly into several different sorts. Of these sorts, a few (four are shown but the exact number is not certain) are closely allied, one is rather less close. These represent the sorts of gibbon. The great apes are much further away, with the chimpanzee closer to the gorilla than to the orang, and man is furthest of all, with no other sort close to him.

It is obvious that there are several natural groups here. First are the actual dot-clusters. Next in size and dispersal over

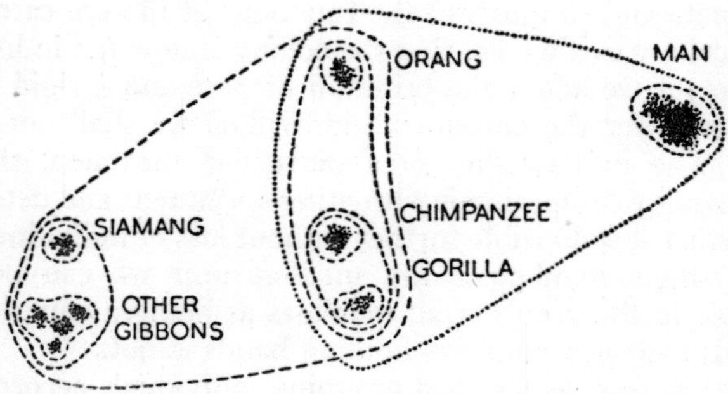

FIG. I. Affinities of the gibbons, apes, and man

the diagram is the group of gibbons other than the siamangs, and the group including the chimpanzee and gorilla. Less compact is the group of great apes. Man is very distinct in some respects. It would be quite possible to combine him and the great apes in one group, leaving the gibbons in another, or, according to which characters are emphasized, to bring together the gibbons and apes, leaving him separate. But it would be quite unnatural to lump the siamang, the gorilla and man in one group.

Now let us fit the hierarchy to these particular forms. Shall we call the main gibbons' stock one genus, the siamang another? If so, should the chimpanzee and gorilla be together in a third, or should they be separate? Are the gibbons as a whole a genus, a subgenus, or family, an order or even a class? The only possible answer is that no one knows, and as far as studying the animals goes, what is important is the recognition of the natural groups, not their names and ranks. But for purposes of reference, of course, it is important that people should agree on ranks and names. The main groups of animals are by consent called phyla. Monkeys of all descriptions clearly are mammals, that is, they form a small group within a well-defined larger one which itself is only one of fourteen within a particular phylum. Obviously, it would be inappropriate to call the gibbons a class, or even an order. They are somewhere down in the family or subfamily region. Simpson actually puts them as the subfamily *Hylobatinae* which with the subfamily *Ponginae* (great apes) and one extinct one makes up the family *Pongidae*. There are eighteen families in the Primates, that is, the rest of the known primates tend to fall into natural groups of about the same degree of internal similarity as the group of gibbons plus great apes. And this group is so homogeneous when compared with such groups as the Mammalia, or the Insecta, that it cannot be placed high up in the hierarchy.

With the exception of the rank of species, which demands and receives far fuller treatment later on, there is nothing more to any rank than this. 'The Order' can be defined only as any natural group which consists of groups of subordinal rank and is contained in a group of superordinal rank—and which it is desired to name. All these rank-names have a purely positional value and no more, and that is their significance, except in so far as they point to a low or high degree of heterogeneity. When the importance of characters can be defined and given numerical values, then the degree of heterogeneity in a group can be made precise, and rank-names may acquire a quantitative significance, but until then they should not be taken too seriously. There is no need to expect everyone to agree on the rank of every group. A particular subfamily may be, in the opinion of one taxonomist, a family, and only a tribe according to someone else. Within limits, it does not matter. In the past, some authors have written about orders and classes as though they were mystic entities in their own right, and the determination of the exact rank of a particular group was made as seriously as if the assemblage concerned were a candidate for entrance to a political society. One still finds some zoogeographers who seem to think that all groups with the same rank-name must have the same status in zoogeography—a very naïve attitude. Rank names above the species are useful provided they are not made to do too much.

Let us now remove the limitation imposed at the beginning of this chapter and admit fossil forms into the classification. Because of the well-known imperfections of the fossil record—the tendency for only animals with hard parts and living particular modes of life to be fossilized at all, the discontinuity of fossil-bearing strata (which are very rarely uniformly fossiliferous throughout their thickness) and the complete absence of fossiliferous rocks from many parts of the geological sequence—many groups of fossils are as isolated from their nearest relatives as are any living animals. But with others we find that, as already mentioned in the case of the Carnivora, there are long series of fossils of which the most recent forms are very distinct and often very specialized, but the oldest are almost indistinguishable one from another. Groups that at present are clear, natural, and easily definable are found to extend back in time losing their peculiar features as they go, and merging into a basal unspecialized group in such a way that classification becomes almost impossible. In fact, the diagram of relations of living forms just discussed (fig. I) is only a cross section at the present time of a solid classification, drawn very diagrammatically in perspective for the carnivora in fig. II.

In this diagram, the vertical dimension represents time, the other two allow for related groups to be displayed as areas, with different sorts of characters in different directions and degrees of difference according to distance.

It will be seen from this diagram that the earliest fossil seals yet discovered are obvious and good seals. In consequence, the suborder *Pinnipedia* which contains the seals, sealions and walruses, remains a distinct group—in fact some authors make it an order. The true bears also (Ursidae) remain a compact group unknown before the Pliocene, but one Miocene genus *Hyaenarctos* is so intermediate in character between the bear group and the closely related dog group that it is uncertain whether it is an unspecialized dog-like bear, or a bear-like dog. *Hemicyon*, another Miocene genus, is in much the same plight. The main stock of the present day carnivores, however, is connected very closely with the Miacidae, as already mentioned, and

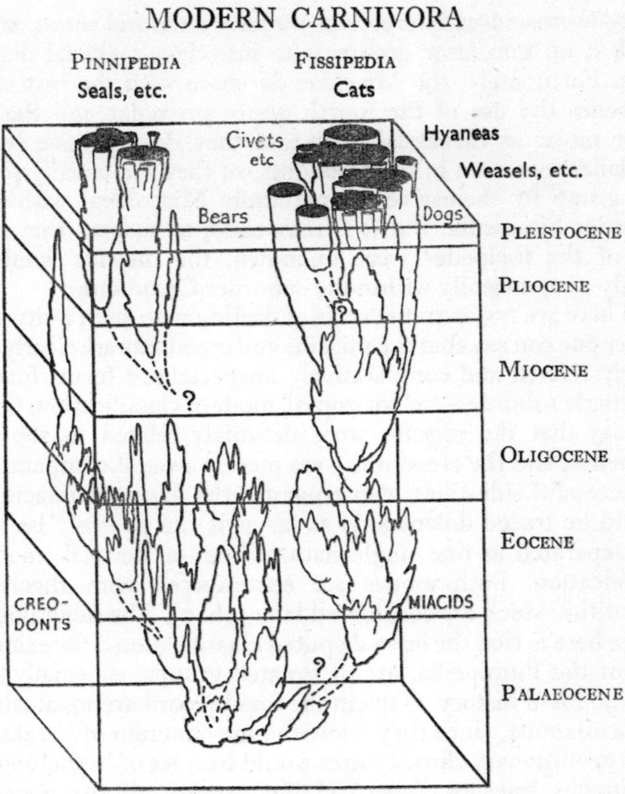

FIG. II. The evolutionary tree of the carnivores

the earliest Miacids are very close to some lines of the Creodont stock, the base of which is almost inseparable from very early groups of different orders of mammals. The main mass of the Carnivora, the suborder *Fissipedia*, has been a clear and obvious natural group since the beginning of the Oligocene period. But in the Eocene it would have been classified as a single family not markedly different from several others then existing, and in the middle of the Palaeocene it might hardly have been given the status of a subfamily. How far back should subordinal distinction be carried, and how can the earliest members be separated from the rest of the forms then existing? In any such lines as the fissipede-miacid-early creodont we have a natural series, and to break it up into large groups is to introduce artificial distinctions. Fortunately, the Miacidae do share with the rest of the fissipedes the use of the fourth upper premolar and the first lower molar as carnassial teeth and they do not have certain specializations seen in the creodonts, so they are usually placed in a group by themselves (superfamily Miacoidea) within the suborder Fissipedia. But it is reasonable to suspect that if the rest of the fissipedes were unknown, the miacids would be merely a superfamily within the suborder Creodonta.

There are two extreme ways of dealing with such a situation. Either one can say that the miacids and creodonts are clearly very closely related and comparatively unspecialized forms forming one single suborder—a 'horizontal' mode of classification. Or one can say that the miacids were definitely related to the later fissipedes, and the creodonts were merely a set of comparatively unsuccessful side-lines, consequently the fissipede-miacid line should be traced downwards as far as it can possibly be taken and separated as one single natural line—a 'vertical' mode of classification. Both modes are encumbered with theoretical difficulties which are considered later (ch. 7). The main point to notice here is that the least-disputed classifications, for example, that of the Pinnipedia, are of isolated groups, especially those with no fossil history. Gaps in the fossil record are invaluable to the taxonomist, since they allow the introduction of breaks into what evolutionary theory states would be a set of branching and continuous lineages if we did but possess all the necessary evidence. In short, the natural system of classification employing mutually exclusive groups is admirable at any one stage of the earth's history, but is adequate for classifying continuous time-series merely because most of them are known only from fragments.

To sum up, it is convenient to distinguish large groups as such from smaller ones, and in consequence a series of rank-names is in use which allows one to estimate the position of any particular group in the natural hierarchy, and thence its approximate degree of homogeneity. An alternative method would be to use only the particular name of each group, and memorize their arrangements. This is quite practicable for the largest groups (down to classes) and would save much unprofitable discussion about the relative ranking of some of them, but it is clearly impracticable for the smaller ones, of which there are many thousands. With the exception of the upper-end rank (which refers to the groups containing all animals, the lower the individual, not being formally recognized as a taxonomic category) and of the species, the ranks have a purely positional value and are defined solely with respect to each other. Furthermore, the distinctions between groups of living animals may disappear when fossil forms are considered, and indeed should do so according to the theory of evolution. Classification into discrete groups is fairly easy at any one time, but is not easy over long periods of time.

CHAPTER IV

NAMES

THE first concern of the taxonomist is to distinguish and describe natural groups of animals. When he has done so, he must name his groups. Classifying and naming are two quite separate and distinct activities, which must not be confused. Names are used in zoology (and botany) as convenient brief labels by means of which taxonomic groups can be referred to quickly and accurately. Once the group *Carnivora* for example has been distinguished, adequately defined, and named, there is no need to list its contents or re-describe it every time one wishes to refer to it. The accepted name is sufficient.

Reference to zoological groups is by words (names), not by mathematical or other symbols, becaues the earlier taxonomists used the current Latin, Greek or vernacular names of particular groups of animals as those symbols of reference to these groups which were the most readily intelligible to people in general. They saw no necessity for breaking away from ordinary names and adopting a special reference-system. Special systems, numerical or otherwise, have been proposed from time to time, but the system of names is too firmly established to be easily discarded, and moreover works very well, on the whole.

Names are arbitrary. It would not matter in the least if all bears were always called men, and vice versa. There is no link between the words *man*, *homo*, *hombre*, etc. and men except by custom. The choice of a name for a group is therefore in the last resort, entirely arbitrary. Names are of very great importance in zoology. One cannot discuss animals at all without having agreed names for the different sorts; and there are certain rules, the International Rules of Nomenclature, which zoologists in general have agreed to keep so that the system of nomenclature shall be internationally understood and always applied in the same way. These rules are of necessity rather complex. In addition there are a series of 'Opinions' given on their meaning or application in particular cases, which together with the rules form a sort of legalistic excrescence on taxonomy, unfortunate but probably unavoidable. As names of groups must be freely used in any discussion of species, a very brief account of them, sufficient only for the purposes of this book, will be given here.

Linnaeus was the first to use consistently the binary or binominal system of nomenclature, and our modern system of animal (not plant) nomenclature is taken to date from the publication of the tenth edition of his *Systema Naturae* (effectively 1st January, 1758). No names published before then (even if binary) have any standing. The *Systema* is an excellent catalogue of the forms of animals known at the time and a very convenient starting point. The system is called binominal because every species is designated by two words, one the name of the genus in which the species is placed and the other a name or epithet applied to that particular species—for example, Linnaeus named the lion *Felis leo*, the tiger *Felis tigris*, and the domestic cat *Felis catus*, all these being species grouped by him within the single genus *Felis*—the cats. Each genus is given a *generic name* which is a single word, always written with an initial capital letter, and which is, or is treated as, a Latin noun. No two genera within the animal kingdom may have the same name (otherwise all sorts of ambiguities would creep into the reference system) but there is no objection to the same name being used for a genus of animals and for one of plants, since confusion between the two is most unlikely. Many years ago Hoffmeister named a genus of worms *Haplotaxis* but on discovering that the same name was used in botanical taxonomy proposed *Phreoryctes* instead. When the International Rules of Nomenclature were accepted, all substitutions of this nature became unnecessary so *Phreoryctes* was abandoned in favour of *Haplotaxis*.

In naming a species, the generic name is always written first. Unless clarity is lost thereby, the generic name is usually abbreviated to its capital letter after it has been given once in full. There are no rules concerning this usage. Sometimes it is convenient to use the first two letters (e.g. *Pt.* for *Ptilinopus*, p. 65). The word that follows it is called by some writers the *trivial name*. Others call it the *specific name*, although many use this term, rather sensibly, for the two names together, since it is the pair that uniquely designates the species; the trivial name must only occur once in each genus, but there is no other restriction on it, and it may be employed very many times. The trivial name is, or is treated as, a Latin word which is an adjective (e.g. *Ursus horribilis*, "bear horrible", the grizzly bear) or a noun in the genitive (e.g. the earthworm *Lumbricus friendi*, named in honour of the Rev. H. Friend) or a noun in apposition (e.g. *Felis leo*, "cat, lion", the lion).It is a very general practice, recommended but not enforced in the Rules, to write the names of genera, species and subspecies in italics, or some other type differing from the rest of the text. In zoology, but not in botany, trivial names are not written with an initial capital letter, except that one can do so if one wishes with trivial names derived from personal names. There is nothing gained by doing so. The many recommendations included in the Rules concerning the choice, formation and spelling of names are not of importance here.

Contributions to the classification and taxonomy of a group may be made by people anywhere in the world, often working independently and seldom with access to all the literature on any single group. Consequently it is not surprising that some words have been proposed as names several times and for different groups, and merely to quote a particular name may be insufficient to specify its application. And further, when a name is first validly published, some indication of its meaning must be given. So to specify a particular name exactly, one must state both it, and the author who published it, and the date when he did so. Thus the full citation for man is '*Homo sapiens* Linnaeus 1758.' The author's name (sometimes abbreviated) and the date follow the specific name immediately. If the species has since been transferred to a different genus, or the generic name has been changed for any other reason, the author's name is placed in parentheses. In a recent check-list of mammals, Linnaeus's large genus *Felis* which included all the cats is split up into several genera, and the lion, for example, is cited as '*Panthera leo* (L.)', the great cats being transferred to the genus *Panthera* established by Oken in 1816; 'L.' is by agreement the standard abbreviation for Linnaeus and no one else, and in the example just given is placed in parentheses because of the change in generic name.

The only category below the species that is expressly provided for in the Rules is the subspecies. The subspecific name is a word, formed in the same way as trivial names, which follows the specific name immediately with, if necessary, its own author and date. For example, the Common Shrew occurs throughout Europe and was named by Linnaeus *Sorex araneus*, the full citation being *Sorex araneus* Linnaeus 1758. The Common Shrews of Britain are rather lighter in colour than those of the continent and have been separated as a distinct subspecies, the full name for which is *Sorex araneus castaneus* Jenyns 1838. The full name of a subspecies is therefore a trinomial. When a species is thus broken up into subspecies, the animals on which the specific name was first bestowed are referred to by using the trivial name also as the subspecific name so that the Common Shrews of Sweden (and the rest of Northern Europe, since these do not differ from the Swedish ones) are *Sorex araneus araneus* Linnaeus 1758. Such subspecies are referred to as the nominate subspecies, or races. It must be emphasized that they are not in any way specially typical or representative of the species. They are merely the forms which happen, through various accidents of history, geography and sometimes politics, to have been named first.

The only category between the species and the genus that is provided for in the Rules is the subgenus. Genera can, if desired, be broken up into subgenera very much as species can be divided into subspecies. The subgeneric name is a word formed and treated in the same way as a generic name and when required is placed in *parentheses* between the generic and trivial names. For example, there are three closely related parrots in Australia which are included in the genus *Polytelis*. In the most recent revision of the parrots one of them is regarded as rather distinct from the other two and is placed by itself in a subgenus, *Spathopterus*. Usually it is sufficient to refer to this species as *Polytelis alexandrae*, but a more precise reference would be *Polytelis* (*Spathopterus*) *alexandrae:* the other species are then referred to as *Polytelis* (*Polytelis*) *swainsoni* and *P.* (*P.*) *anthopeplus*. Some taxonomists feel that the subgenus is not to be recommended. Its use makes names cumbrous, and it is a standing temptation to the next reviewer of the group to treat it as a genus—but when this happens the generic names of all the species in all but the nominate subgenus must be changed. Such authors would refer to *species-groups* within a genus which requires subdivision. Such groups can be used by the specialist to indicate the fine degrees of affinity within a genus, and cause no nomenclatorial troubles. The name of a species is a binomial. When a subgeneric and subspecific name are added, a very cumbersome quadrinomial results.

When the same name has been inadvertently published for two or more different forms it is a *homonym*. Different names published for the same form are called *synonyms*. The Rules lay down that the correct name for any form is that which conforms to the binomial system and is the first validly published in or since the *Systema Naturae*. This is the famous 'Law of Priority.' 'Validly published' means that the name must have been printed, and must conform to certain other conditions. If the same name has been used more than once its later homonymous applications must be dropped for ever. A synonym, however, can be revived if necessary. If some reviewer considers that species *A.b.* and *C.d.* are really the same he must use the older name (*A.b.*, say) for both, and the newer name must be listed as a synonym of the older. But if subsequent work shows that *C.d.* is really different, its status can be restored. Or if *A.b.* is discovered to be a homonym of some yet earlier name, it must be dropped for ever, and if *C.d.* is a synonym of it and is the oldest of all the synonyms, the species must now be called *C.d.* When a well-known generic name has been suppressed in favour of another, it is often convenient to mention it when the new name is used, since it will be more readily recognized. If it is given in brackets after the new name it must be in the form *A* (=*B*) *c.*, otherwise it will be given the status of a subgenus, which will be entirely wrong.

Names must be permanently rejected as homonyms, and can be suppressed if considered to be later synonyms. It is important to note that (once published) they cannot be rejected because they are inappropriate. For example, many species of birds have been named on the evidence of skins bought from traders who often, intentionally or unintentionally, gave a false locality for them. The small parrot *Vini peruviana* was supposed to come from Peru. In fact, it is a native of Tahiti, and when this was discovered, its name was changed to *Vini taheitana* as being more appropriate. This substitution is no longer allowed, and it must be called *V. peruviana*. In Linneaus's day the scientific name of an animal might often be its vernacular name latinized, or a very brief description of it, and it was not unreasonable to correct inappropriate names. In modern usage, the name need not be a description, and in its literal meaning may be wholly wrong when applied to a particular species. It is only a means of reference to that species—a designation, not a description.

Two other points require mention. The first is that the Rules for names of families and subfamilies state that they must be derived from the name of one of the genera to which they refer and must end in *-idae* and *-inae* respectively (e.g. the dogs, *Canis*, whence the dog family *Canidae*). These and other standardized endings in general use have been referred to above (p. 31). The second is that the Rules do not give any guidance on the naming of varieties. Specimens that differ from the normal are frequently named as *varieties* and it is customary to give to a variety a name which is formed on the same lines as a trivial or subspecific name and which is quoted after the name of the species concerned but never directly following it. A black variety might be referred to as "*A.b.* var. *nigra*" with an author's name and date. Various synonyms or partial synonyms for variety are in use, such as *aberratio* or *mutatio*. Many different things may be named as varieties—an individual variation (even caused by disease), a seasonal form, a geographical subspecies, or even a good but not recognized species. In consequence the term is not used in the best-worked groups, such as birds, where the correct status of such forms is known. Individual variants are not dignified by a special name in such groups because all animals differ in some respects; to name a few only is to direct disproportionate attention to those few, and to name all individuals is impossible. In less well-worked groups (the majority) the custom of naming varieties is somewhat more respectable because many good species were first recognized and named as varieties; because they were named, they were sought after and studied and enough information was obtained for their true status to be established. Unfortunately in some groups, notably insects and molluscs, some authors still give a formal name to every striking individual variant. This practice is not to be recommended. It produces a huge mass of names and is a strong incentive to mere variety-hunting.

CHAPTER V

THE POLYTYPIC SPECIES

IN the series of obligate taxonomic categories given in the preceding chapter the species is the lowest. If we consider the animals that ordinary European people knew best in Linnaeus's day, they were (as now for many) the larger mammals, some birds, some butterflies, and possibly a few others. Now, although there are many different varieties of horses, sheep, dogs, pigeons, and other domesticated animals, no one has much difficulty in recognizing even an extreme variant, such as the Shetland pony, bulldog or pouter as a horse, dog, or pigeon respectively. Again, several different sorts of hawk have been well known from very ancient times in Europe because of the popularity of hawking. Intermediates between the merlin, the hobby, the peregrine falcon, and so on, are exceedingly rare, or non-existent. The hybrid between the horse and donkey is indeed known but only under rather artificial conditions and it is sterile. Linnaeus and his predecessors could therefore readily assume that people would know several examples of *sorts* of animals that were clearly distinct in nature but resembled one another so strongly that they could be grouped together. Not only could individual birds be recognized as kestrels, say, but the resemblance of kestrels in general to peregrines, hobbies and others meant that they could be spoken of collectively as falcons in opposition to other birds of prey such as vultures.

When Linnaeus introduced the generic name *Falco* for many of the birds of prey, and distinguished *Falco tinnunculus*, the kestrel, *F. subbuteo*, the hobby, *F. rusticolus*, the gyrfalcon and various others, he was in effect merely formalizing in the then universal language of Europe what had already been done in the vernaculars. The genus represents a familiar general sort of animal or plant, the species a particular sort within it. And, individual variants and domestic breeds apart, the species was the lowest category. As with the falcons, so with the raven, the rook, the carrion and hoodie crows and the jackdaw. All are crows of a sort. Any large black bird in Europe (with a few interesting exceptions discussed in detail later, p. 95) can be assigned readily to one or other of these sorts or species of crow. In general, there are no intermediates between good species, provided one restricts oneself to a fairly small region of the earth. Linnaeus accepted and proclaimed the doctrine that there were as many species of animals and plants as had been created in the beginning; but we may be sure that this brilliant man would have noticed any divergence between this dogma and his own experience in the field, and would have commented on it, at least to his pupils. This point is of the utmost importance, since it meant that species were accepted as essentially static although with considerable individual variation, and differing quite clearly from each other. Between the most closely similar species there was a gap. Normally, no intermediates occurred, or if they did they were few, and sterile or nearly so, not truly linking the two species.

Here, then, we have the ideas that the species is the smallest group above the individual, and that each species is clearly separate from those most closely resembling it. These ideas have been of enormous importance in taxonomy; indeed it would not be unfair to say that they characterized the species-concept nearly until the present day. One can go even further. Since the majority of forms of animals were known, and still are known, almost entirely from dead specimens, another criterion widely recognized, namely that species reproduce after their kind, or 'breed true', cannot be used, nor can behaviour of any description; and the majority of species of animals are described purely on the basis of their anatomical features. Now, what actually happens when a new species is described?

Suppose that there arrives at one of the great museums from some remote region, a number of bottles of preserved animals and that a few earthworms are among them. These will be picked out, and sent off to one of the very few taxonomists interested in the Class Oligochaeta, together with whatever data on season and locality the collector happens to have supplied. The taxonomist consults the standard works on Oligochaeta plus all the numerous taxonomic papers scattered about in all sorts of scientific periodicals, and fairly readily determines the families and genera to which the specimens belong. Then in each genus he tracks down the descriptions of all the species known and carefully compares his specimens with them. Some specimens, no doubt, will agree exactly with one or other description or so closely in all but a few minor characters that it is obviously the description that is slightly at fault, not the identification. If based originally on only a few specimens the description may well give as a constant character for the species as a whole what is in fact only a usual one; or it may even be actually wrong on some small point. If the taxonomist can do so, he will compare his specimens with the actual specimens that original descriptions were based on, since they are the final authority, so to speak, and he can check both his own specimens and the details of the original descriptions against them.

Now suppose that with one exception all of his specimens agree sufficiently well with published descriptions; he will label them with the appropriate species-names, and that concludes his work on them as far as identification goes. But the exception requires further investigation. Is it a new species, and if so should it be placed in a new genus? If it is not a new species does it represent a geographical race of some known species? Is he sure it is not merely a rather juvenile individual of a well-known species, with some characters only partly developed? Can it be an individual that has lost a few segments anteriorly and regenerated a different number, thereby appearing to have genitalia displaced from their normal segments? Or is it a well-marked variety, occurring with normal individuals, like white-headed blackbirds or albino rabbits among normal ones? If, as in this example, he has only a single specimen, only his experience of similar earthworms can help him. All he has to go on is his general knowledge of what immature or regenerating earthworms look like, what sort of variation does occur in individuals of related species, and the actual morphological characters of this particular specimen.

Fortunately it is true that individuals of any species have a huge number of characters all which help to make up the general characteristic appearance associated with that species, and it is unlikely that all these will alter out of recognition in any one individual. Consequently the taxonomist who has studied a group knows approximately what sort of variation to expect, and he is on the whole far more likely to give a reliable judgement on a particular specimen than are other people. This is the basis of the pragmatic definition of the species, approved by Darwin, and others since—that a species is a species if a competent systematist says it is. (It is equally true that other people are dissatisfied if a taxonomist assures them that a particular specimen belongs to a particular species because he knows it does on the basis of his experience but can hardly point to a single diagnostic character to confirm his identification; but this situation does not often occur, and usually if only one could find time, diagnostic characters or groups of characters could eventually be worked out.)

On the basis of his experience, then, the taxonomist can say that his exceptional specimen is e.g., merely an unpigmented individual of a well-known species, or, that it differs so much from all species so far known that it must be regarded as a new one. In the latter case, he publishes a new name supported by a sufficient description to enable others to recognize this new species. The specimen, being unique, must be the *type* of the new species. This means only that the new name is for ever associated with this particular specimen; it does not mean that this specimen is typical of the species. Discovery of more individuals may prove that the original description is wrong in almost every particular. But if there is still no doubt that it is a distinct species, the name given to the first specimen must still be used, provided it is not a homonym, with the original describer's name. If it is decided that it was only an extreme variation of a species already known, then the new name must be dropped because it is a synonym of the older one. But if, years after this has been done, further work shows that the specimen did represent a good species after all, then that first specimen still carries the name with it, and the species must be known by that name.

Because of this restriction to purely morphological comparisons, let us describe species recognized in this way as *morphological species*, more briefly *morphospecies*. It will be seen at once that all (or very nearly all) species founded upon fossils will of necessity be morphospecies. (But in palaeontology, transformation in time must be taken into account, and this produces other difficulties, discussed later.)

From the description just given of what happens when a

morphological species is described, it is obvious that the morphological species is a convenient and natural pigeon-hole for one or more actual specimens, plus all those specimens that may eventually turn up which resemble the original sample sufficiently closely. If many intermediates between the original and some other sample are discovered, the two may be combined under the older name. Exactly the same procedure holds for establishing any other natural group of any rank whatever—genera, orders or phyla, with the sole exception that the types of genera are species not specimens, and ranks higher than the family do not have types at all. The only peculiarity about the morphological species is that it is the lowest taxonomic rank. (Individuals that depart more or less from the normal are of course known, and can be classified as varieties, aberrations or monstrosities according to their degree of divergence but they are of course exceptional.)

We can sum up the characteristics of any morphological species as follows:

(1) It is static, with no reference to changes in space or time.
(2) It is monotypic, containing only a group of individuals approximating fairly closely to a single norm of variation.
(3) It is the lowest taxonomic rank (except in so far as aberrant individuals may have been given varietal names).
(4) It is defined entirely on morphological characters.
(5) Each species is almost always clearly separated from its nearest relatives.

In fact, the morphological species was, and still is, a very useful device for cataloguing the immense variety of animals that had begun to stream into museums in Linnaeus's day, and still continues to do so. The first task of all for any taxonomist is to catalogue the enormous numbers of sorts of animals. Nearly always he has only preserved specimens or their equivalents (casts, descriptions, illustrations etc.) to work with. And it must be remembered that three-quarters at least of the million species of animals known are known taxonomically only from preserved material, and are classified, from pure necessity, into morphological species. The morphological species is still, and must be, the most useful device of the taxonomist, because of our almost complete ignorance of the majority of species. Yet one cannot be satisfied with it, and the reasons for this state of affairs must be examined.

As a result of the zoological and botanical exploration of the world, vast numbers of new species were discovered. The taxonomist, almost overwhelmed by more and more discoveries, was forced to spend most of his time in devising keys and compiling catalogues of known forms. But as certain groups of animals which for various reasons were popular, came to be well studied, great difficulties arose over the separation of species At the present day, the groups of birds (the class *Aves*), is probably the best-worked group of animals as far as taxonomy goes, and it was the ornithologist who was first influenced in his taxonomic practice by these difficulties. It was discovered that very frequently the zoological exploration of a previously unworked region brought to light forms which were both geographically and morphologically intermediate between well-known and closely allied but distinct species living in nearby regions, so that a geographical series of very closely allied 'species' could be constructed.

For example, there is only one species of wren in Britain, namely *Troglodytes troglodytes* (L.), and it was formerly considered to be spread all over Britain and the nearby islands, as well as through Europe. But when specimens from some of the smaller islands were examined they were found to differ constantly from those of the mainland. The St. Kilda wren is a comparatively large bird, rather greyer brown above and paler below. The barring is heavier on the wings, and more extensive on the back. The Faroes wren is like the St. Kilda wren in size but intermediate between it and the mainland wren in colour-characters. The Iceland wren is even larger, but a dark brown above. The Shetland wren is as large as the St. Kilda wren, but darker than the mainland form. The Hebridean wren is like the Shetland wren in colour but the underparts are more buff, and the barring is less heavy. In size it is more like the mainland forms. Wrens from Skye are intermediate in character between the Hebridean and the mainland forms. Lastly, it was found that birds from western Scotland tended to be more uniformly dark on the upper parts, although individual specimens from elsewhere on the British mainland might be as dark—in fact,

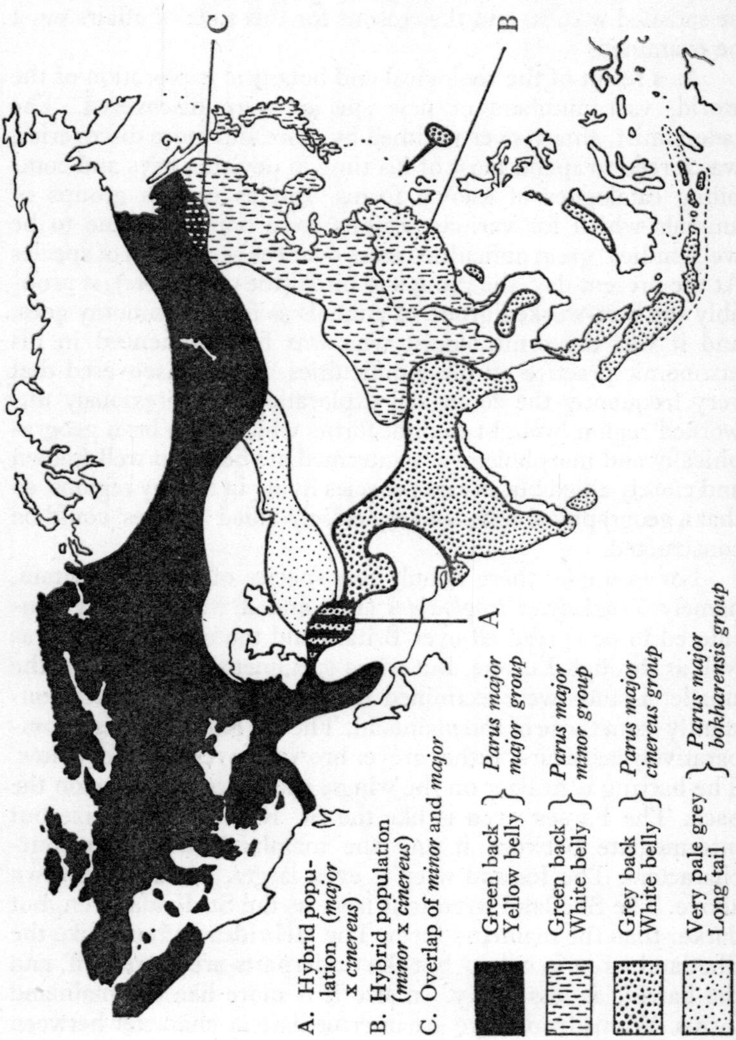

FIG. III. Distribution and variation in the Great Tit

although they differed, there was some overlap of variation.

The wren was named binominally (as *Motacilla troglodytes*) by Linnaeus. The St. Kilda wrens were not recognized taxonomically until 1884 and were then described as a separate species, *Troglodytes hirtensis*. For very similar reasons, the Islay shrew, which is closely allied to the British common shrew but differs in size and colour pattern was originally named as a species, *Sorex grantii*, Barrett-Hamilton and Hinton 1913, but is now regarded as a geographical race, or subspecies, *S. araneus grantii* of the common shrew. The musk-shrew of the Scilly Isles, *Crocidura cassiteridum* Hinton, is so similar to one of the common Continental species that it too might be considered only a subspecies.

The wren varies noticeably even within the British islands. Other birds do not; their characters are much the same anywhere in these islands, and indeed, may stay almost constant from Ireland to Japan. For example, the Great Tit (*Parus major*) the range of which is shown in Fig. III, is a common, conspicuous and popular bird of Europe, which extends almost without change from Ireland through northern Europe, Russia and central Siberia to the Pacific, reaching the coast of the Sea of Okhotsk (between Kamschatka and northernmost Japan). The birds of the British Archipelago have a thicker beak and a slightly darker green back, and are distinguished as the subspecies *Parus major newtoni*, but the rest are almost uniform in their characters except for a little variation in size, so that the nominate race (that is, the one that was first named) extends as *Parus major major* from Brittany to Okhotsk. In this form the back is dull green, the belly yellow.

In the south of Europe, however, there is some geographical variation. The birds of Spain, Portugal and North Africa are very like *P.m.major*, but the yellow of the belly is more vivid and the white areas along the outer tail-feathers are reduced (*P.m. excelsus*). The birds of the Balearics, Cyprus, Crete and Greece (*P.m. aphrodite*) have a more greyish back and a paler yellow belly than has *P.m.major*. The white patches on the wing-feathers are intermediate in average extent between those of *major* and of *excelsus*. The birds of Corsica and Sardinia agree with *P.m. aphrodite*, but as they are slightly greyer on the flanks as well, they have been separated as *P.m. corsus*. Those of Palestine (*P.m. terraesanctae*) are somewhat paler beneath and more yellowish on the back than any of the preceding, and this tendency is carried further in the birds of Persia and North Mesopotamia (*P.m. blanfordi*).

All these forms, unless isolated by the sea, tend to intergrade. Indeed, the Palestine forms are so intermediate between the North European and the Persian that were it not for their geographical position as an offshoot from the main area of the species they might hardly be given a separate name. Where the sea-barrier is slight, as between Kent and northern France, intermediates may be found in spite of it. The Kentish birds do display some similarities to those of Northern France. The populations isolated on the Mediterranean islands are more distinct, but all the forms mentioned are extremely similar to *P.m. major*, and one can regard them only as local races.

However, in India there is a bird obviously very closely related indeed to the European Great Tit, but which differs sharply from it in having a grey, not green back, and a white, not yellow belly. In Japan and Manchuria is another which does have a green back, but the belly is white, not yellow. It might well be thought that the various races of *Parus major* in Europe differ (by comparison with these other forms) in such slight characters that if they are regarded as belonging all to one species (which is obviously correct), then the Indian and Japanese forms must be assigned to two other species. Even using the criteria of the morphological species there is nothing else one can do. But the situation is not really as simple as that. Among the forms that were first recognized taxonomically were *Parus m. major* Linnaeus 1758, *Parus m. cinereus* Viellot 1818 (from Java, with a grey back like the Indian bird), and *P.m. minor* Temminck and Schlegel 1850 (Japan). These were naturally described as species. Later, many other races were discovered, including some that appeared intermediate.

As already mentioned, the Persian birds (*P.m. blanfordi*) are paler than those of Europe. The Indian birds have a grey back and white belly. In Khorassan, to the south-east of the Caspian Sea, there are populations of a large *Parus* which are white beneath and grey above, but with a light blotch of green on the back which is extraordinarily variable, being absent in many individuals. This population connects the paler races of the undoubted *Parus major* of Europe with the large number of grey-backed races including *cinereus* that extend from just east of it south-eastwards through India, Burma, Malaya, Sumatra, Java and the lesser Sunda Isles. It is regarded as a hybrid population; grey-backed and green-backed birds have entered this region from east and west respectively, and have successfully interbred here.

Again, throughout much of South and Central China, and along a strip extending down the coast of Indo-China, exists a form, *P.m. commixtus*, which resembles the grey-backed forms, especially that of Java, except in that the tail is even greyer, and that the back may be more or less tinted with green. In some ways it is transitional between the grey-backed forms of India, Burma and Malaya, and the green-backed (but white-bellied) forms of Japan (*minor*), Korea, Manchuria, Inner Mongolia, Tibet and the mountains of north-east Burma and extreme western South China. Where it approaches these forms geographically there is some intergradation of characters. This form may or may not be hybrid, but it certainly is intermediate between the Oriental grey-backed forms like *cinereus* and green-backed forms like *minor*; it is not possible to maintain that they belong to separate species. Grey-backed forms resembling the Indian and Javan birds are found on Hainan and small islands north of Formosa (Ishigaki, Iriomote). Apparently the species does not occur on Formosa and Sakhalin.

However, even greater difficulties arise. The white-bellied, green-backed forms of the Extreme East extend northwards from Japan and Manchuria up as far as the Amur River, which is the northern boundary of Manchuria. Here they meet *Parus major major* which, as mentioned above, stretches from Brittany to Okhotsk. In the upper Amur valley both *P.m. major* and the northernmost white-bellied form *P.m. minor* coexist and breed without interbreeding. They remain distinct just like perfectly good species. Yet *P.m. minor* is connected *via P.m. commixtus* of China and the grey-backed forms of Indo-China and Malaya, with the Indian forms which hybridize in Khorassan with the group of subspecies including *P.m. major*.

In the region of Lake Balkash live populations which are exceedingly pale and have very long tails. Four subspecies are distinguished, which are, in order from west to east, *bokharensis*, the smallest, *ferghanensis*, which is darkest, *iliensis* large and pale, and *turkestanicus* with a distinctly larger beak. Two nesting birds of *P.m. bokharensis* have been recorded inside the territory of *P.m. intermedius*, the hybrid race of Khorassan, just to the south-west. And in the north-east it seems that *turkestanicus* meets *P.m. major* in the region of the Altai mountains without interbreeding. It would almost seem that these very pale forms with long tails may behave as a good species where they meet or overlap both *P.m. major* and *P.m. intermedius*.

The situation is summed up in figure III, in which only the main groups of subspecies are shown. (A quite conservative review lists thirty-three subspecies!) Much more information from Central Asia is required before definite conclusions can be reached, but it seems that while the pale forms from Turkestan may perhaps be regarded as a separate species (which by the rules of nomenclature should be called *Parus bokharensis*) the rest are a series of morphologically interconnected forms, the end-members of which overlap in the Amur region and behave there as good species. Alternatively, one might consider that as the hybrid *P.m. intermedius* is confined to Khorassan, it represents so small a region of hybridization that all the Oriental forms from *P.m. minor* in the Amur Valley south to Java and west to Afghanistan could be called a separate species, *P. minor*, overlapping with *P. major* in the north without interbreeding, and in the west with only a small zone of hybridization. Whatever course is taken, it is obvious that none of the species delimited can be brought within a single morphological definition, because the characters vary, well beyond the limits of individual variation, from population to population.

If two forms meet geographically and are found to intergrade smoothly (primary hybridization) there is no reason to regard them as anything more than local differentiations of the same species. If they meet and give rise to a great outburst of variation (secondary hybridization) then the meeting may be comparatively recent and there must be considerable genetical differences between the two parent forms. Whether they are called separate species or not depends on the width of the hybrid zone and other considerations. If they meet and overlap with no, or only the rarest, hybridization, then there is no doubt that they are behaving like good species.

But this means that it is very difficult to determine the status of forms restricted to islands (whether actual islands in the sea, or islands of forest in a sea of steppe, or oases in a desert does not matter). Only the degree of morphological resemblance can be a guide, and if a closely related group of forms consists entirely of insular geographical representatives, the situation is desperate. The honey-eating parrot *Trichoglossus haematodus* and its relatives provide an excellent example.

Trichoglossus haematodus and its geographical representatives range over the Australasian region from just east of Java through New Guinea, the Solomons and New Hebrides to New Caledonia, with replacing forms in north-western and in eastern Australia, and one remarkable form on Ponape in the Caroline Islands far to the north-east. Most of these birds have a black head with blue streaks, a yellow-green collar, and the rest of the upper parts green. Beneath, the breast is red or yellow (with or without black bars on the ends of the feathers), and the belly is blackish green. The form in Celebes differs from all the others in having the cheeks and throat red, as well as the breast, a red patch on the collar, yellow feathers in the arm-pits instead of red ones, and no yellow stripe on the underside of the wings. It is usually listed as a distinct species, *Trichoglossus ornatus*. The form on Ponape is the most distinctive of all. It is a dark red-brown all over, with hints of black bars on some of the feathers. The tail is honey-coloured, and beneath the wings there is only black—no yellow stripe or red arm-pit patch. It is separated as *T. rubiginosus*.

But some subspecies of *T. haematodus* are almost as distinctive as *T. ornatus*. The north-western Australian form has a bright blue head, a bright orange collar with a flush of dark blue on the back just behind it, an orange breast with no bars, and a dead black belly. The form from Flores is almost entirely green, but shows traces of a yellow breast and of blue streaks on the head.

Others are less distinctive, but still quite noticeably peculiar. For example, the form from the island of Biak, north of the great north-western bay of New Guinea, is conspicuous not only for the extremely heavy barring (black with a beautiful purple gloss) of the breast, but also for a huge yellow patch comprising the collar and much of the upper back. Yet other forms are so alike that it is necessary to examine several specimens of each to appreciate the differences. The birds of north-western, northern and south-eastern New Guinea, of the vast arc of islands stretching from the Bismarcks to the New Hebrides, and of New Caledonia, are all remarkably similar. Like the south European forms of the Great Tit, they are slight local modifications of the same stock, and must be classified as geographical subspecies.

None of these various forms overlap anywhere. In New Guinea three adjacent forms are known to intergrade smoothly, as one would expect, and the other two races on this island probably do so. But all these are very closely related anyway. The very distinctive north-western and eastern Australian forms are well separated by unsuitable country. All the other highly distinctive forms are on islands. As with the Great Tit, one can distinguish fairly easily groups of closely allied races, and again these groups are geographically rather compact. Closely related forms are not found scattered here and there among others that differ from them considerably. But there is nothing to help one in deciding where specific limits shall be drawn. There are twenty-four replacing forms to be arranged, and they could be lumped together into a single species or split into approximately eighteen species with equal ease, according to the opinions of the reviewer about the degree of morphological difference which should be taken to indicate specific difference in this group. The two species most closely related to the *T. haematodus* group overlap with some members of it and behave as good species, but they are both very distinctive—as peculiar as *Trichoglossus ornatus*. Had either been very like one of the *haematodus* group, one would know that very slight differences might indicate specific status in these coconut lories; as it is, the significance of slight distinctions remains undecided.

Comparatively recently, Australia was far less arid than it is today, and forests extended far into the interior. The arid period seems to have set in very suddenly. Its effect was to confine forest-living animals of all sorts to several refuges where rainfall was sufficient to maintain the sort of habitat necessary for them. The principal refuges are in the extreme south-west, in the south-east (a large one), in the north-east (probably the Atherton tablelands), and in a small area around Darwin in the Northern Territory. Some birds were able to survive only in the huge

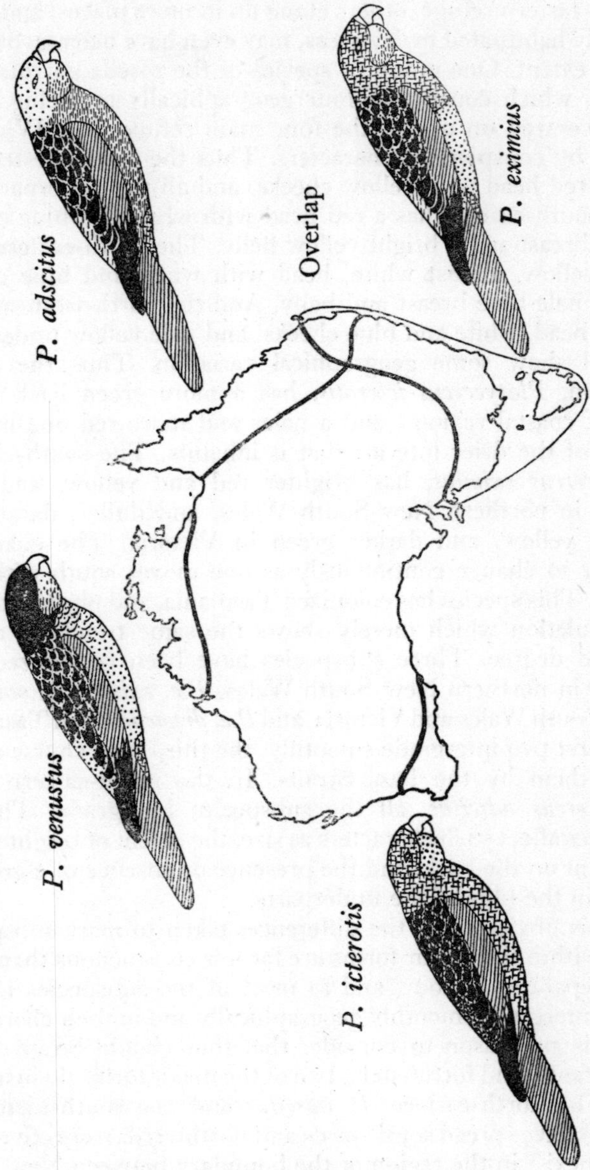

FIG. IV. Geographical variation in Rosellas

south-eastern refuge, others clung on in more places, and some, already habituated to dry areas, may even have been favoured to some extent. One group of species of the rosella parrots exists today, which consists of four geographically replacing forms, each centred on one of the four main refuges (fig. IV). They differ by conspicuous characters. Thus the south-western one has a red head with yellow cheeks, and all the underparts red. The south-eastern has a red head with white and blue cheeks, a red breast and a bright yellow belly. The north-eastern has a pale yellow, almost white, head with white and blue cheeks, and a pale blue breast and belly. And the north-western has a black head, white and blue cheeks, and pale yellow underparts.

All show some geographical variation. Thus the south-western, *Platycercus icterotis*, has a more green back in the wetter coastal regions, and a paler and more red one in those parts of the drier interior that it inhabits. The south-eastern, *Platycercus eximius*, has brighter red and yellow, and paler green in northern New South Wales, and duller, darker red, duller yellow, and darker green in Victoria. The characters appear to change continuously as one moves south within its range. This species has colonized Tasmania, and produced there a population which merely shows the same trends to an enhanced degree. Three subspecies have been recognized, *P.e. cecilae* in northern New South Wales, *P.e. eximius* in southern New South Wales and Victoria, and *P.e. diemenensis* in Tasmania. The first two intergrade smoothly, the third is of course cut off from them by the Bass Straits. In the north-eastern form, *Platycercus adscitus*, all the subspecies intergrade. The differences affect such characters as size, the extent of bright yellow pigment on the back, and the presence or absence of a greenish tinge in the blue of the underparts.

It is obvious that the differences taken to mark subspecific rank within these four forms are far less conspicuous than those that separate the four, and as most of the subspecies in each form intergrade smoothly geographically and in their characters, there is no reason to consider that they should be given any other rank. And fortunately, two of the major forms do just overlap. The north-eastern *P. adscitus* and the south-eastern *P. eximius* have spread southwards and northwards respectively and now coexist in the region of the boundary between New South

Wales and Queensland. Here they appear to behave like good species. There is no zone of secondary hybridization, still less a smooth intergradation. And since the difference between these two species are of the same sort as those between these and the other two, it is reasonable to give all four specific status. The isolated Tasmanian form of *P. eximius* is so similar to the mainland ones that it is reasonably given only subspecific status.

But in a closely related species, the very beautiful Crimson Rosella, *Platycercus elegans*, the situation is not so easily understood. This is a larger bird than the four just mentioned and prefers on the whole the denser forests; it is not a bird of open parkland. It is crimson all over except for blue cheeks, and the blue areas on wings and tail and the black scalloping on the back common to all the rosellas. It occurs mainly in the forested regions of eastern and south-eastern Australia, and there are closely related forms, probably subspecies, in south Australia.

In Tasmania also there is a large form of *Platycercus*, which in general build and habits is obviously far closer to the Crimson Rosella than to the four smaller species, but it is dark green above and yellow-green below, with occasional tinges of red on the under tail coverts, and a narrow red band on the forehead, just above the beak. As far as colour-characters go, it is almost less like the Crimson Rosella than the latter is like the four smaller Rosellas. It is always considered a separate species—*Platycercus caledonicus*. But the apparently large gap between it and the Crimson Rosella diminishes when one finds that the juvenile plumage of the latter is also predominantly green, that when red appears in juvenile *P. elegans* it tends to do so first on the head and under tail coverts, and that records of female Crimson Rosellas breeding while still in the green plumage have been published. It is quite possible that the Tasmanian Rosella is derived from a Crimson Rosella which retained juvenile plumage throughout its life. In fact a character highly conspicuous to human eyes is not necessarily of greater importance than a less conspicuous one in assessing specific status. Again, the Tasmanian and Crimson Rosellas are isolated from each other in nature, so their status cannot be determined with certainty.

The Golden Whistler or Thickhead *Pachycephala pectoralis* is a very pretty songbird which occurs in one form or another from Java to Fiji and Australia. The female is an inconspicuous brownish or brownish-green bird. The male usually has a black cap, a white throat, a conspicuous black band separating the white of the throat from the golden yellow of the breast and belly, and an olive green back and wings. About eighty geographical races have been described. Some differ only very slightly; sometimes only the females differ, usually the males are distinct. In a few races the male has the same plumage as the female, in one the female is almost cock-feathered. This enormous array of geographically replacing forms has a curious rather V-shaped distribution, stretching from the Northern Moluccas and Java southwards and eastwards to Australia and then northwards and eastwards *via* the New Hebrides and Solomon Islands to the Bismark Archipelago, and eastwards to Fiji. New Guinea lies in the V, separating its arms, and although rich in species of this genus has no member of *Pachycephala pectoralis*, except that one Australian subspecies reaches the south-east coast. One of the New Guinea forms, *Pachycephala soror*, is obviously the representative in New Guinea of *pectoralis*, but one subspecies of *pectoralis* overlaps its range and does not intergrade with it. Consequently *P. soror* cannot be included in *P. pectoralis*. Yet there is little doubt that if this overlap did not occur, *soror* would be regarded as a subspecies of *pectoralis*, and a not extremely peculiar one at that. It is hardly as distinctive as *Pachycephala pectoralis feminina* from Rennell I. in the Solomons, in which the cock is hen-feathered, or *P.p. sanfordi* from Malaita in which the cock has wholly golden yellow underparts from the chin to the tail, with no black breast band, or *P.p. graeffii* from some islands of Fiji, in which the black breast band is absent and the forehead is yellow.

In fact, *P. pectoralis* plus its representative extends all over the Australo-Papuan and adjacent regions, and the curious V-shaped distribution of *pectoralis* itself is probably not a genuine zoogeographical fact. Yet if one decides that *pectoralis* is to be broken up into several species, then exactly the same difficulties arise as did in the case of the coconut lory, *Trichoglossus haematodus*—where shall the boundaries be drawn? There is only one overlap to help us. Apart from that, all the populations of *pectoralis* are isolated from one another. And so many show intermediate characters.

In this example, two undoubtedly good species (*P. pectoralis* and *soror*) are distinguishable by plumage differences which are not very great, yet since the genetical importance of some of the more conspicuous differences is unknown, and since intermediates are plentiful, the overlap does not help us to arrange the isolated forms into species. At first sight a cock *P. pectoralis* with the plumage of a hen is so different from one in the usual male dress, that one is inclined to put it instantly into a separate species. But the juvenile plumage of both sexes is very similar to that of the adult female. The change-over to hen-feathering in the male may not be so very profound after all.

Even in groups of closely related species, the same character may obviously not have the same significance every time it occurs. For example, there is a genus of very beautiful small fruit-pigeons in the Pacific, from Java eastwards, which have very diverse colour-patterns. Approximately ninety geographical races are recognized, and these are grouped into about thirty species, several of which are isolated geographical representatives of others and might be combined with them by an extreme 'lumper'. The greatest number of actually overlapping species is twelve (in New Guinea) but two or three is more usual.

The juveniles, and the females when they differ from the males, are principally green. The males vary enormously. Some are green with a huge patch of red, white or yellow on the chest (*Ptilinopus rivoli*, *solomonensis*, and *viridis*), others have a grey or green head and a deep purple patch on the belly (*Ptilinopus hyogastra*, *granulifrons* and *naina*). One is green with a grey head, a large black blotch on the back of the neck, an orange chin-stripe, and deep red under tail coverts (*Pt. melanospila*). Others have much more complex patterns; for example *Pt. superbus* has a large red cap on the head, dull red shoulders, a bronze-green back, green wings with large blue-black spots, pale grey throat and breast, a large blue-black band across the belly, and the rest of the underparts creamy white with large green blotches on the flanks and legs.

In almost all the species, excepting only those that are confined to one island or small island-cluster and those that are nomadic over large areas, almost every character of the colour-pattern varies geographically to some extent, so that characters constant through one species may be of only subspecific value elsewhere. A grey head is a specific character of *Ptilinopus melanospila*, *hyogastra* and *granulifrons*, which are probably quite closely related. But it also appears in some forms of several other species, sometimes becoming nearly white. A yellow-bronze head is characteristic of *Pt. perlatus* of New Guinea; yet in its very close relative, *Pt. ornatus* one subspecies (in the west of New Guinea) has a dark red cap. A red cap is a specific character of *Pt. wallacei* and *superbus* and is nearly always found in the very closely allied series of geographically replacing species *Pt. coronulatus*, *regina*, *porphyraceus*, *purpuratus* and their allies. But in this series it has been lost or reduced nearly to vanishing point at least four times—in the Marquesas (*Pt. dupetithouarsii*), in several subspecies of *Pt. purpuratus* east of Samoa, in the southern Solomons (*Pt. richardsii*), and in some islands of the Banda Sea between Celebes and north-western Australia (subspecies of *Pt. regina*).

Two pairs of species are particularly interesting. One consists of *Pt. hyogastra* on Halmahera and Batjan and *Pt. granulifrons* on the nearby island of Obi Major. Both these species are green with a grey head, deep purple belly-patch, and yellow under tail coverts. They are almost indistinguishable—except that *granulifrons* has a large walnut-like outgrowth on its forehead. *Pt. iozonus* (New Guinea) and *insolitus* (Bismarck Archipelago) are also closely related. They are green with an orange belly-patch, white or yellow edges to the undertail coverts and some grey spots on the wing. In some subspecies of *iozonus* there is a purple patch on the wing, and there may be no grey band across the end of the tail feathers. In *insolitus* there is more orange on the belly, no grey on the chin, paler (less bluish) grey spots on the wings, a large grey band on the tail, and wide yellow edges to the under tail coverts. All these characters taken together would probably earn for *insolitus* the rank of a rather well-marked subspecies of *iozonus* were it not that *insolitus* also has a large orange knob on its forehead—an excrescence similar to that in *granulifrons*. Such growths, developed from the inflated region around the nostrils, are almost always specific characters elsewhere in the pigeons. Thus of the four large barking pigeons (*Ducula*) that over-lap in the Solomons, the Grey Pigeon (*D. pistrinaria*) and Chestnut-bellied Pigeon (*D. brenchleyi*) have no knob, the Pacific Pigeon (*D. pacifica*) has a large black one, and the Red-knobbed Pigeon (*D. rubricera*), as its name implies, has a large red one. Because of the walnut of *Pt. granulifrons* and the orange knob of *Pt. insolitus*, their possessors are always given specific status. *Pt. insolitus* was even placed in a genus (*Oedirhinus*), by itself. But they are geographical representatives of their closest allies, and in the case of *Pt. granulifrons* the resemblance to *Pt. hyogastra* is so close that one wonders whether the two should not be listed as two geographical subspecies of a single species.

In one case, subspecific variation in *Ptilinopus* has produced convergence good enough to puzzle even experienced ornithologists. *Ptilinopus rivoli* is a green pigeon with a large white breastplate, *Pt. solomonensis* is green with a large yellow one. But in one subspecies of each both breastplate colours are found coexisting. *Pt. rivoli bellus*, from high ground in New Guinea, has a white breastplate with a distinct yellow tinge in the middle. *Pt. solomonensis speciosus*, from some small islands in Geelvink Bay (the great bay in the north-west of New Guinea) has a yellow breastplate which has acquired a distinct white edging.

In this example convergence is by the acquisition of characters. Vaurie has described a more complex example of convergence in two rock nuthatches (*Sitta neumayer* and *Sitta tephronota*) of south-eastern Europe and western Asia. These overlap widely in Persia, and where they overlap there is a well-marked difference in the wing length, and in the size and shape of the beak. Also, a dark stripe running from the beak to the shoulder is very poorly developed in *S. neumayer* and particularly well marked in *S. tephronota*. But where the two do not overlap the differences are far less marked and examples of *S. neumayer* from Dalmatia and *S. tephronota* from the eastern part of its range (Fergana) are remarkably alike, so much so that until comparatively recently there was great confusion over the dividing up of the various populations into species.

It seems, then, that although at any one locality species are almost always clearly distinguishable, they may vary and intergrade geographically to such an extent that it may be impossible to provide a simple and accurate description fitting all the geographical forms, or to provide any criteria for distinguishing between species and mere geographical races. This situation does not occur only in birds. It is probably true to say that geographical variation causes considerable difficulty in every group of animals which is well enough worked for it to be discoverable. It is well known in mammals, reptiles, amphibia, fishes, many sorts of insects and other arthropods, some earthworms, some echinoderms, and a few other groups, and will probably appear wherever it is looked for, except in those species that are distributed so readily that any local varieties that arise are continually swamped by immigrants. But so many groups are so badly known that not enough material is available to allow one to sort out what differences between the known specimens are due to individual variation, what to geographical, and what are interspecific differences.

When the morphological species was universally accepted, a vast number of species of birds was described. But very many of these were closely allied, and could be arranged as series of geographical representatives. As can be seen from the examples given above, their boundaries with respect to one another might be ill-defined or even indeterminable. Many turned out to be only local races; others were more distinct and might be called species, yet if they were isolated from each other in nature one could not be sure, since the significance of character-differences is very hard to estimate, as the examples just given clearly show. Some highly variable geographical series contain forms more different from one another than from others that overlap with the series and behave as good species (e.g. *Pachycephala pectoralis* and *soror*). Some geographically isolated and therefore non-intergrading populations are more like parts of an intergrading series than its two ends are. For example, in the Great Tit, the birds of Great Britain and of Hainan hardly differ from those of the nearest portions of the nearby continents, except in having a thicker beak, whereas the extremes of the series (*P.m. major* from Okhotsk and *P.m.minor*, or even *P.m. major* and *blanfordi*, *P.m. minor* and *cinereus*, if the series is split) differ in considerably more than this. Obviously the British and Hainan populations are only local races and must be included in the main series. Actual intergradation is not the sole criterion for a subspecies.

Some workers felt that these geographical series were so different from the Linnaean morphological and monotypic species that the same word could not be used to refer to both.

Richard Kleinschmidt was one of the first to understand the difference, and he used the term *Formenkreis* ("array of forms" —implicitly geographical) for the geographical series. More recently, Rensch has substituted the term *Rassenkreis* (array of races), for such series, in contrast to *Art* (species) for forms which do not break up into geographical races. Some geographical series contain members which apparently should be given specific rank. These he termed *Artenkreise*. With further work, it was found that the Rassenkreis was more common than the Art. As Dr. Mayr has remarked, "it was realized that *Rassenkreise* were not a special and rare kind of species, but rather that the majority of species were polytypic. The workers in ornithology have become so used to the fact that the species is a group concept that most of them have now given up the term *Rassenkreis*, since this is merely another name for the species in its modern, revised meaning."

The modern concept of the species is therefore a considerable extension of the old monotypic and morphological one. The lowest taxonomic category is no longer the species but the geographical race, or better, the geographically defined population. Most populations differ to some degree in some characters. The members of all populations show individual variation. The first task of the systematist now is to discover what are the characteristics of the various populations that make up a species. The *variety* as a taxonomic category is abandoned, since it has been used to contain distinctive specimens irrespective of whether they were individual genetic variants, seasonal forms, representatives from geographically and morphologically distinct populations, or even closely related species. One now requires from each principal area inhabited by a particular form a good series of specimens which will enable one to sort out individual, sexual or seasonal variations from those that characterize the population of that area. If it is found to differ sufficiently from all others the population can be named. (Ornithologists usually employ some variant of the rule that seventy-five per cent of the specimens from a population must be distinguishable from all those from the most similar population for the first to require taxonomic recognition.) The rank *subspecies*, formerly employed for any striking variety, is used for such geographically definable, taxonomically distinct populations within a species. Mayr has proposed as suitable for international use, the term *superspecies* instead of *Artenkreis* for arrays containing populations or groups of populations that are distinct enough to be given specific rank. When, as with isolated forms, it is a matter of great difficulty to decide whether certain populations are to be ranked as species or subspecies, he has suggested the term *semispecies* to mark their intermediate nature. Other workers have produced many other terms, but most are superfluous.

The use of these terms can be illustrated from the examples given above. If a *slight* geographical overlap may be permitted, the rosella parrots form two superspecies, which are referred to by the oldest specific name available, thus:

Platycercus elegans superspecies

P. elegans	Eastern, south-eastern and southern Australia.
P. caledonicus	Tasmania.

Platycercus eximius superspecies

P. eximius	South-eastern Australia and Tasmania.
P. adscitus	North-eastern Australia.
P. venustus	North-western Australia.
P. icterotis	South-western Australia.

If the forms in a superspecies must be completely separate geographically, then the *eximius* superspecies must be broken in two, so that *adscitus* and *eximius*, which do overlap slightly, are separated. If this is done, the two superspecies so formed can be included in the *eximius* species-group, to be contrasted with the *elegans* species-group which contains the large rosellas in a single superspecies. The *species-group* is a very convenient rank within the genus. Its use involves no new names, and it can be ignored by the non-specialist.

A complete taxonomic citation of the rosellas would give all their subspecies as well, together with author, date, and references to the original description and important synonyms, as in the following examples (from which synonyms have been omitted).

Platycercus eximius cecilae Mathews.

P. e. cecilae, Mathews 1911, Novitates Zoologicae vol. 18. p. 14, new name for *P. splendidus* Gould 1846, preoccupied by *Psittacus splendidus* Shaw 1792.

Type locality: Darling Downs, Queensland.

Range: South Queensland, northern New South Wales, intergrading with the following.

Platycercus eximius eximius (Shaw).

Psittacus eximius Shaw 1792, Naturalist's Miscellany vol. 3, plate 93.

Type locality: 'New Holland'=New South Wales according to G. M. Mathews.

Range: New South Wales, Victoria, South Australia.

Platycercus eximius diemenensis North.

Platycercus diemenensis North 1911, Australian Museum Special Catalogue no. 1, vol. 3 pt. 2, p. 128.

Type locality: Tasmania.

Range: Tasmania.

Such a list as this has enormous conveniences. First, it greatly reduces the number of specific names. For example, the great ornithologist Salvadori, using the morphological species, recognized sixty-four forms of the fruit-pigeons, *Ptilinopus*, all of which were of specific rank. Since his revision twenty-seven more forms have been recognized, but the number of specific names has dropped to about twenty-nine, since so many are now regarded as subspecies only. A further grouping would produce about fourteen superspecies in about eight species-groups. It is much easier to refer to "*Pt. melanospila* and its subspecies" than to refer to *Pt. melanospila*, *bangueyensis*, *talautensis*, *xanthorrhoa*, *aurescentior*, *pelingensis*, *chrysorrhoa*, *margaretha*, *massoptera* and *melanauchen* as a group of closely related species, leaving the reader to discover for himself that they are more like each other than any other form of *Ptilinopus*, all geographically representative, and a few actually intergrading. The assistance given to the reader by the use of the trinomial is so great that it is often a good practice to reduce semispecies to subspecies, so that their names indicate their nearest relatives. It may be objected that perhaps the population in question really is specifically distinct. But, as already pointed out, if it is wholly isolated, its status cannot be finally determined, and perhaps it is really only subspecifically distinct. Where its status cannot be conclusively demonstrated one may as well use the more convenient nomenclature for a population.

Secondly, such a brigading into polytypic species and superspecies is of great help in zoogeographical and faunistic studies. It makes clear the relative status of the various forms. For example, in the Society Isles and Tuamotu Group (S. Pacific) there are six named forms of *Ptilinopus*, in Fiji and Samoa there are only five (excluding 'Chrysoena'), in the Solomons eleven, in New Guinea and its faunistically associated archipelagos, thirty-four. All these were or would be called species, formerly. In actual fact, all those from the first region are now placed as very similar subspecies of *Pt. purpuratus*, whereas of the Fijian and Samoan forms, three are subspecies of *Pt. porphyraceus*, itself a representative of *purpuratus*, while two are subspecies of the highly distinctive and endemic *Pt. perousii*, which overlaps *porphyraceus* in range. Faunistically, Fiji and Samoa, with fewer named forms, are actually richer than the Tuamotu group and Society Isles in species of *Ptilinopus*. The thirty-four New Guinea forms reduce to thirteen species, of which two are in a single superspecies. Those from the Solomons reduce to five species, and the derivative nature of the Solomons fauna is beautifully demonstrated by the fact that three of these species also occur in New Guinea, and the other two are members of a superspecies with a representative in New Guinea. The use of the polytypic species and superspecies in this genus emphasizes the real richness of New Guinea, the comparative poverty of the Solomons, and the affinities of the Solomons forms. Obviously the polytypic species and superspecies must be used as faunistic units, not the old monotypic species, in zoogeographical studies.

The polytypic species has been equally useful in evolutionary studies, but this subject requires special study (Ch. VIII, p. 130).

To sum up, the morphological species, the lowest taxonomic category, static, monotypic, and without reference to variation in space or time, is still of necessity used to classify the greater part of the animal kingdom. Where sufficient material is available to enable us to study geographical variation, the geographically definable population becomes the unit of study. Such units are grouped into subspecies, these into (polytypic) species, and these, if necessary into superspecies. The polytypic species does take account of variation in space, and does not conform to a single norm of variation. Geographically representative species are not always clearly delimited either geographically or morphologically. The introduction of the polytypic species marks a great step forward, which can only be taken in sufficiently well-worked groups.

CHAPTER VI

THE BIOLOGICAL SPECIES

THE polytypic species, then, has one great disadvantage. The place of geographically completely isolated forms within the hierarchy superspecies-semispecies-species-subspecies cannot be conclusively demonstrated. Two terms introduced by Mayr are useful in discussing their status. "Two forms or species are *sympatric*, if they occur together, that is if their areas of distribution overlap or coincide. Two forms (or species) are *allopatric*, if they do not occur together, that is if they exclude each other geographically." We can determine by observation the specific or merely varietal status of partially or completely sympatric forms, but can only guess at the status of wholly allopatric ones. As was shown in the preceding chapter, the exact significance of morphological characters may be very hard indeed to determine and may vary greatly even between closely related species. There was little difficulty with the old morphological species. Different forms were different species unless they actually intergraded when they were variants of the same species. And the species of a small region, which are sympatric, usually differ quite sharply from one another in morphology, physiology and behaviour.

Very many 'good' species are known—that is species with clear-cut morphological characters, that breed true and do not hybridize with their nearest relatives when they meet them in the wild. In Britain, familiar examples are the Great Tit, Blue Tit, Coal Tit, Marsh Tit and Willow Tit (*Parus major*, *caeruleus*, *ater*, *palustris* and *atricapillus*) or the Song Thrush, the Mistle Thrush and the Blackbird (*Turdus ericetorum*, *viscivorus* and *merula*) or the Large Cabbage White, the Small Cabbage White and the Green-veined White butterflies (*Pieris brassicae*, *rapae* and *napi*). The members of these groups differ not only in their morphological characters but in every other way—in voice (if any), preferred foods, nesting-sites and nest material (in the birds), eggs, period before hatching, etc. For example, *Pieris brassicae* lays its eggs in batches, *rapae* and *napi* lay theirs singly.

The eggs of all three are skittle-shaped with fifteen to seventeen longitudinal ribs in *brassicae*, eleven to thirteen in *rapae* and about fourteen in *napi*. The caterpillars are gregarious in the first-mentioned species, much less so in the second, solitary in the third. Those of *brassicae* are regularly blotched with black on a yellow ground above, and green below. Those of *rapae* (when full grown) are dark green with a faint yellowish line down the back and a line of small yellow spots on each side near the legs. Those of *napi* are dark green with a darker green line down the back, a faint yellowish one on each side near the legs, and whitish grey underparts. *P. rapae* and *brassicae* feed on many sorts of cultivated cruciferous plants, and a few wild ones, *napi* almost entirely on wild ones. *P. brassicae* is abundant throughout the British Archipelago except in the Shetlands, and is usually found over cultivated ground. *P. rapae* is also very abundant, but is absent from Shetland and the Hebrides. *P. napi* is seldom seen over cultivated ground, preferring lanes, hedges, and rough ground where the foodplants grow. It does not go as far north as the others, being absent from the Shetlands, the Orkneys, the Outer and some Inner Hebrides, and most of northernmost Scotland. Conspicuous colour variants in *brassicae* and *rapae* are very rare, less so in *napi*, in the Irish populations of which a yellow form has occurred several times. There is no need to list the specific characters of the adults, which will be found in any book on British butterflies or better, in the summer, on the animals themselves.

Such examples could be multiplied almost indefinitely. Closely related sympatric species usually differ in a host of ways, and there is seldom any difficulty in separating them, particularly if living specimens can be studied. With some species behavioural characters may be the most easily appreciated, as every bird-watcher knows. Flight, stance, and above all voice supply invaluable characters for recognition. Courting behaviour is often important since many species have highly characteristic rituals. The courting 'dance' of slugs, for example, is often highly specific. But ordinary day-to-day behaviour may be just as useful. Anopheline mosquitoes usually rest with the tip of the abdomen raised so that the abdomen and the proboscis are in one straight line, while Culicines usually rest with the abdomen parallel to the surface on which they are resting or even with the tip pointing down to it. Juvenile individuals of the red and black slugs (*Arion ater* and *rufus*) can be easily distinguished from other species (e.g. the small and very common *Arion hortensis*) by a curious rocking movement they display when touched.

But there are certain difficulties. Although so many species are clearly distinct morphologically, physiologically and behaviourally as well as genetically, some are not. There is every degree of morphological distinctiveness within groups of closely related sympatric species. Although general biological distinctiveness is usually associated with clear-cut morphological peculiarities, the association is not invariable. One can construct a series beginning with 'good' species distinct in every way and ending with some which are morphologically indistinguishable. This series is epitomized in the following examples:

(i) The earthworms *Allolobophora chlorotica* and *A. caliginosa*, good species with constant and clear morphological differences.
(ii) The chaffinch and brambling (*Fringilla coelebs* and *F. montifringilla*) good species with very slight anatomical differences, but strong behavioural ones.
(iii) The snails *Cepaea nemoralis* and *C. hortensis*, good species with well marked but inconstant differences.
(iv) The tree-creepers *Certhia familiaris* and *C. brachydactyla* with very slight but very constant morphological differences.
(v) The mosquitoes of the *Anopheles maculipennis* group, morphologically indistinguishable in the adults but separable by characters of the eggs.
(vi) The fruit flies *Drosophila pseudoobscura* and *D. persimilis*, apparently morphologically indistinguishable in all stages.

(i) The very common earthworms *Allolobophora caliginosa* and *A. chlorotica* are clearly distinct morphologically. Both are originally European but have been spread by man over all the temperate parts of the world. Both are very common in cultivated soil. Neither comes to the surface at night. *A. chlorotica* is common among decaying vegetation, for example under old overturned turves, but not in compost heaps. *A. caliginosa* is more common in garden earth. *A. chlorotica* is usually dull green or yellow, seldom pink; *A. caliginosa* is whitish, pinkish or brown. In *A. chlorotica*, the glandular band (the clitellum) which secretes the cocoon extends like a saddle over segments 28 or 29 to 37, and on each of segments 31, 33 and 35, at each edge of the clitellum is a curious tubercle. In *A. caliginosa* the clitellum extends over segments 26, 27 or 28 to 34 or 35, and on each side two large tubercles extend from segments 31 and 33 towards each other, meeting or actually fusing on segment 32. When *A. chlorotica* is touched it usually curls up into a circle, like the caterpillar of a moth, whereas *A. caliginosa* wriggles and tries to escape. When irritated, *A. chlorotica* emits from pores on each segment a considerable quantity of the body-cavity fluid, which is yellow with suspended matter. *A. caliginosa* does not. Both can be bred easily in captivity. Like all earthworms they are hermaphrodite, but in them cross-fertilization takes place. No hybrids can be produced, and none have been observed in nature. These are good species in every way.

(ii) The chaffinch and the brambling (*Fringilla coelebs* and *F. montifringilla*) are also good species. In both, the male and female plumages differ, but far more obviously in the chaffinch than in the brambling. They are illustrated in most books on British birds. The brambling is only a winter migrant in Britain (although nesting has been reported very rarely), and breeds further north, whereas the chaffinch breeds throughout most of Britain. But apart from their plumage patterns and a slight difference in size, there is very little indeed to distinguish the two species anatomically. Their plumage and behavioural differences are clear-cut, obvious and constant, their anatomica differences are not obvious at all.

(iii) The two snails *Cepaea nemoralis* and *C. hortensis* (they have no genuine vernacular names) are very common in western Europe, and *C. hortensis* extends to Iceland and Labrador, ranging further north than does *C. nemoralis*. They are easily recognized since the shell is yellow, pink or brown, rarely lilac or other colours, and is striped with any number of bands up to five (very rarely more) any of which may fuse with adjacent ones, and any (including all) may be absent. It is very rare indeed for a colony of either species to include only one banding-pattern; usually a large number are present, that is, each colony is highly polymorphic. Colonies containing both species are not uncommon, but on the whole the two tend to have rather different distributions in any region.

The shell characters of these two species, then, are as obvious as the plumage characters of the chaffinch and brambling, but unfortunately, they are not constant. All the possible banding-patterns have been recorded for *C. nemoralis*, and most for *C. hortensis*. The various shell colours are almost all known from both. Yet the frequency of these variations varies considerably. In Britain yellow is a far more common shell-colour in *C. hortensis* than in *C. nemoralis*. The variety with a single black band down the centre of the whorls is common in *C. nemoralis* and very rare in *C. hortensis*. But in Europe this banding pattern is not uncommon in *C. hortensis* as well. *C. nemoralis* has a larger shell than *C. hortensis* but there is considerable overlap. In most parts of their ranges, the strong lip that strengthens the mouth of the adult shell is black in *C. nemoralis* and white in *C. hortensis*, and usually this can be taken as a characteristic difference between the two species. But white-lipped *nemoralis* and dark-lipped *hortensis* do occur as individual variants, and in Ireland and in the Pyrenees white-lipped *nemoralis* are so common that this character is useless. In fact, although most colonies can be sorted readily on the basis of a *combination* of shell characters, only two certain diagnostic characters are known, namely, the number and shape of the branches of a mucus-gland associated with the genitalia, and, better, the shape of the dart, an extraordinary weapon secreted by the animal and used to prod the other partner during courtship. This is a crystalline rod with four flanges running along it at right angles to each other, so that a cross-section would look like a four-pointed star. In *C. nemoralis* the edges of the flanges are simple, but in *hortensis* each flange forks into two smaller ones near the free edge. Hybrids between the two species have been produced and show darts of every possible degree of intermediacy between these two types; such darts have hardly ever been observed with certainty in wild animals. There is no doubt that the two species are good ones, yet their most obvious characters, with very few exceptions are not constant.

(iv) The European tree creeper is a small, very attractive little bird, pale below and rather wren-like above, that climbs up tree-trunks and large branches, picking off insects with its rather long curved beak. Several geographical races are known, which vary in colour and size. Many travellers have noticed that the call of the tree creeper in Britain is much weaker than in France. The ornithologist Brehm, after a careful study of these birds came to the conclusion that there were in fact two species in Europe. One is pretty well confined to mountainous districts on the Continent of Europe but in Britain, where it alone is found it is a bird of lowlands. This is the form that was named by Linnaeus and must still be called *Certhia familiaris* L. The common lowland form of the Continent Brehm called *Certhia brachydactyla*. The differences between the two forms are so slight that the specific status of *brachydactyla* was not accepted for many years. *Certhia familiaris* is a purer white below and a less dark brown above, with a shorter beak, a distinctly longer claw on the hind-toe, and a softer voice and less piping call note. But the differences, although constant, are mostly slight and nearly as slight as those distinguishing the different geographical subspecies of each form. Yet the two do show constant differences in morphology, voice, and habitat-selection, and do not interbreed in nature.

Here, apparently, we have two good species with differential characters which although constant are very far from obvious; indeed they were regarded for very many years as quite trivial.

(v) Next in this series of good species arranged with decreasing morphological difference comes an example which has been worked out because of its medical importance. The mosquito *Anopheles maculipennis* is a carrier of the protozoan parasites which cause malaria in man, and is widespread in Europe and the Near East. It is fairly readily distinguished from other species of the genus. A careful study of the numerous races that have been described by taxonomists, plus the breeding of large numbers in the laboratory has shown that this apparently well-defined species is in fact composed of at least five and almost certainly six species which hardly differ in the adult stage. The best diagnostic morphological characters are found to be the colour and arrangement of the eggs, which are deposited as little floating clusters, but behavioural characters are numerous and very useful. The species are *A. sacharowi*, *maculipennis*, *messeae*, *melanoon*, and a pair of very closely related forms, *atroparvus* and *labranchiae* which may be subspecies or distinct species. Several show geographical variation.

A. sacharowi is a clearly marked species living in the eastern

Mediterranean and the Near East. The eggs and larvae are remarkably tolerant of salt water. The adults will only mate in captivity if they are allowed a large cage in which to swarm. This species is an important malaria-carrier. *A. maculipennis* is very widespread in Europe. It has very slight tolerance for salt water, and in captivity requires a very large outdoor cage in which to swarm and mate. It is not, or only very rarely, a malaria-carrier. *A. messeae* is found in the valleys of the larger rivers of Continental Europe. It has very low tolerance for salt water, and is not (or only rarely) a malaria-carrier. *A. melanoon* is a Mediterranean species, with a tolerance for salt water slightly higher than in *maculipennis* and *messeae*. One subspecies has not so far been induced to mate in captivity. It is not a malaria-carrier.

The *labranchiae-atroparvus* group contains the common malaria-carrying European mosquitoes. There is much geographical variation, but the various forms seem to fall fairly clearly into two main groups which are almost allopatric. The *labranchiae* group is found in Italy, Spain, North Africa, and some of the Mediterranean islands. The *atroparvus* group is widespread in Central Europe. There seems to be an overlap in central Italy, yet no intermediate forms are known from this region, which suggests that the two groups are specifically separable. One of the most interesting differential characters is that while *atroparvus* will mate in a very small cage without swarming, *labranchiae* needs a larger cage and must swarm. Hybrids have been obtained in the laboratory, and the males are sterile. The absence of hybrids in nature and their low viability when artificially produced strongly suggest that these two groups are separate in nature. The striking difference in behaviour is of less certain significance since it has been found that in certain mosquitoes of the genus *Culex* an equally striking character, namely the ability or inability of the females to lay eggs without first taking a meal of blood, has a very simple genetical basis.

There is no doubt that the medical importance of the *Anopheles maculipennis* complex has caused it to be investigated with particular care. One wonders how many other 'species' at present known only morphologically are in fact similar complexes. As long as the forms in a group are separated solely by means of morphological characters, such complexes will almost certainly remain uninvestigated.

(vi) Good species that are separable only with difficulty on morphological evidence are usually referred to as *sibling species*. An even more extreme example than the members of the *A. maculipennis* complex is afforded by two species of fruit-flies, *Drosophila pseudoobscura* and *D. persimilis*, frequently referred to as *D. pseudoobscura* race A and B. The group of the true (two-winged) flies, the order *Diptera*, is a huge one which has attracted comparatively few workers, so that most of its known species are morphological, and probably vast numbers remain to be discovered. When it was found that many of the little fruit-flies, *Drosophila*, were extremely suitable as material for genetical experiments, many workers began to study them, to breed them in large quantities, and to discover that there were more species than had been described by systematists. Often they had to turn taxonomist to describe the material they were working on. Because of this increased interest in the genus, especially in America, well over fifty new species have been described and it has been realized that perhaps several hundred more remain to be discovered in the less explored parts of the world.

A very dark species of *Drosophila* was described by Fallén in 1823 from Sweden, as *D. obscura*. In 1921 a dark species was found on the Pacific coast of North America which was referred to this species. But in 1929 it was discovered that the American and north European forms could in fact be readily distinguished by numerous behavioural and structural characters, including the chromosomes (see p. 179), and the American form was separated as *D. pseudoobscura*. (Since then over fifteen new species in the *obscura* species-group alone have been recognized in America, and eight so far in Europe.) But in 1929 it was also discovered that *D. pseudoobscura* consists of two distinct strains or races which are readily distinguishable by their chromosomes. In males, the Y chromosome (the distinctively male chromosome) is J-shaped in one strain, V-shaped in the other. Furthermore, when individuals of these two races were mated in the laboratory, all their male progeny were sterile.

These two races have been the subject of much study. It has been shown that not only are the male hybrids sterile, but the female hybrids, when crossed with males of either race produce less vigorous offspring than do females of either race mated with males of their own race. It seems certain that the hybrids are far less vigorous than pure-bred forms, and could not persist for long in the wild. Even in the laboratory, if a choice is possible males pair with females of their own race. The two races overlap widely in their distribution. Large numbers of wild individuals have been studied, but no hybrids have been found. The races differ in their ecological preferences, one preferring a cooler, more humid and oceanic type of climate, the other a warmer and more continental one. There is no doubt that in their genetics, breeding biology, ecology and many physiological features they behave as good species. Yet they are so alike that for many years they were believed to be identical in all external morphology. By far the best character to use for separating them was the shape of the male's Y-chromosome.

More recently, intensive study has brought to light certain slight differences. On the front legs of the males are a number of stout bristles arranged to form two combs. The average number of bristles in these combs differs in the two forms, as does the shape and size of the wing. A special 'wing index' was devised which is obtained by multiplying the area of the wing (in mm^2) by the cube of the wing length (measured in mm.), which could be used to distinguish the two races. The difference in wing shape and size is correlated with a difference in the rate of wing-beat. A slight structural difference in the genitalia has also been observed. But it is almost impossible to distinguish the two forms from dried specimens such as are usual in museums.

For a long time these two forms were referred to as *Drosophila pseudoobscura* race A and race B. But to do this is to obscure their biological status, and accordingly in 1944 Dobzhansky and Epling named race B as *Drosophila persimilis* giving an admirably reasoned account of the circumstances which led them to do so. As they remark, "It is certain that if any kind of structural difference had been known *D. pseudoobscura* and *D. persimilis*, they would have been classed as species from the start." Some of the differences between these two forms are summarized below.

D. pseudoobscura	*D. persimilis*
Y chromosome J-shaped.	Y chromosome V-shaped.
At 25°C lays more eggs than *D. persimilis*	At 14°C lays more eggs than *D. pseudoobscura.*

D. pseudoobscura	*D. persimilis*
Survives starvation better.	Not so well.
Prefers more continental climates. Very widespread in N. America.	Prefers more oceanic climates; confined to western N. America.
Wing index 45.7–62.8.	68. 8–76.2.

It is obvious, then, that between good species there may exist any degree of morphological difference down to practically nothing at all. Exactly the same situation is found in plants. Should populations which will not interbreed where they coexist in nature but which are indistinguishable on any morphological criterion be called species? Some taxonomists have claimed that forms cannot be called species unless they are recognizable on museum material. Others have felt that to abandon all other considerations and concentrate entirely on the presence or absence of interbreeding is to abandon the natural system of classification for an artificial or special one, which should not be done.

The first objection is the result of taking the morphological species too seriously. If all taxonomists were to agree that only forms with more than a certain degree of morphological distinctiveness could be called species a consistent scheme of classification might be produced, yet its value would not be great. Estimates of the 'degree of difference' tend to be very subjective and to vary from one worker to the next. The catalogue so produced would have to list some perfectly good species as merely subspecies of others if those others showed very considerable geographical variation. It would abandon the advantages of the polytypic species. But in any case there is no reason why the techniques commonly employed in museums up to now should be all-sufficient for detecting the limits of species. One might as well ask all chemists to use only the apparatus known to John Dalton, since he was able with it to produce the atomic theory. And there is no reason why the museum taxonomist should not put together all his specimens of the *Anopheles maculipennis* group, for example, in one drawer, labelling them "*A. maculipennis* aggregate," until diagnostic characters that he can use have been found.

The second objection is much more serious at first sight. If classification is to be by resemblance in all characters, as explained in Chapter I, then surely the fact that two forms are, as far as can be ascertained, identical in all the numerous characters of their extremely complex anatomy should suffice for their classification in one species, even if they do not interbreed. But there are two good reasons why such forms should not be considered conspecific.

In the first place, such forms, *sibling species* as they are usually called, are only the extremes of a series, as has been shown above. Examples of every degree of intermediacy between the most extreme sibling species and the most obviously 'good' species can be found, sometimes—as in *Drosophila*—even within the limits of a single subgenus or species-group. Nowhere in the series is there an obvious gap; nowhere can a natural limit be set to sibling as against 'good' species—indeed, it is noticeable that the expert in any group will maintain that he can distinguish even difficult species quite easily and therefore that there are very few siblings in his group. A distinguished ornithologist will refer to the snails *Cepaea nemoralis* and *hortensis* as siblings; a practised conchologist will remark that they are obviously distinct, far more so than some he could name, and that only inexperience causes them to be regarded as siblings. There is no criterion, and no hope of obtaining one. We must reconcile ourselves to the fact of nature that although morphological and genetical distinctiveness almost always go together, sometimes the morphological fails when the genetical doesn't.

The second reason is simply that in fact morphological distinctiveness *by itself* has never been a sufficient reason for granting specific status. It has always been subordinated whenever possible to the criterion of interbreeding. No one has ever proposed that males and females should be separated into distinct species, yet they always differ in at least the genitalia and often in scores of other characters. If individuals, both male and female but all with the general characters of the human male were known, and others with those of the human female, then unless they all interbred freely they would certainly be regarded as comprising two distinct species. Men and women are far more easily distinguished than *Drosophila pseudoobscura* and *D. persimilis*. In some groups of animals the males and females are so remarkably distinct that only the chance finding of mating pairs will allow them to be sorted out. It is rumoured in entomological circles that until recently two subfamilies of a certain family were recognized. But when it was noticed that all the species in one were described from males only, and all those in the other from females only, suspicions were aroused. I cannot vouch for the truth of this story, but I can for its high inherent probability. In many groups of the animal kingdom males and females are extraordinarily distinct and must be identified from separate keys, since they have hardly a useful specific character in common. Some female parasitic Crustacea, for example, are mere egg-bags with a digestive system attached, while the males, which are free-swimming and hunt for the females, have locomotory appendages, sensory organs, a well-developed nervous system, and other features necessary for their active existence.

Again, the structural changes that take place during a single life history may be extraordinarily profound. It could be maintained with considerable truth that there is more difference between the caterpillar of the Cabbage White butterfly, for example, and the adult than between adults of the white butterflies already discussed (*P. brassicae*, *rapae* and *napi*). Indeed there is more difference between the grub of such a fly as the housefly or *Drosophila*, and the adult than there is between the adult flies and insects belonging to different Orders. The phylum *Echinodermata* (starfishes, sea-urchins etc.) provides even better examples. The larvae of starfish are minute transparent little animals floating in the sea, swimming by the beating of innumerable microscopical cilia, and feeding on very minute animals swept into the mouth by feeding currents created by other cilia. Each is bilaterally symmetrical like ourselves, that is, it has a left and right side, head and tail, back and front (actually an upper and under side). There is little resemblance between this and the large, complex, radially symmetrical starfish, crawling along the bottom by means of its marvellous system of water-tubes and suckers, pulling the shells of oysters, mussels, and similar molluscs apart by main force, and devouring the animals inside. In many features larval starfish resembles larval mussels more than they do adult starfish; yet once their extraordinary transformation is seen, each kind of larva can be matched with its adult form even though at first the two seem whole phyla apart.

The family *Chironomidae* contains a large number of very

gnat-like flies with aquatic or semi-aquatic caterpillars, some of which are coloured bright red with haemoglobin, the pigment of blood. Some of these are the familiar 'bloodworms' of water-butts and ponds, which swim with a characteristic wrapping and unwrapping motion. In some species the genitalia develop precociously in the larvae, which can then reproduce without ever transforming to the adult fly. This phenomenon is called *paedogenesis* (that is, reproduction in the juvenile) and is quite widespread in the animal kingdom. Anyone who comes across a sexually mature larva of one of the paedogenetic forms, and recognizes it as an adult, will be forced, if he takes morphological criteria too seriously to put it into a different order of insects from the normal adults. Of course, no one would dream of doing so. Its resemblance to the larvae of the species of *Chironomus* would be amply sufficient to indicate its real relationships.

In short, not the most bizarre differences of structure can suffice to separate even specifically two specimens that can be shown to exemplify changes in the same life history. One can go even further. The workers of many species of ants, bees and wasps differ conspicuously from the queens and drones. They are not stages in the life history of a queen or drone, and they never reproduce. But because they, as well as queens and drones, can appear among the offspring of a single queen mated to a single drone, they must be included with them in the same species. Genetically, they are dead ends, blind alleys, contributing nothing to the next generation—but that does not matter. Whatever variation is found among the progeny of a single pair of individuals of a species must be regarded as part of the variation within that species. Blue-eyed and brown-eyed men: workers, queens and drones: larval and adult starfish: if the progeny in each case of a single conspecific pair, are also conspecific. Species breed true, not in the sense that the progeny is homogeneous, but that the same sort of heterogeneity is produced from generation to generation. The natural classification holds of course in all comparisons of similar stages—males with males, for examples, or larvae with larvae.

The morphological criterion has been useful only because it points to genetic difference. Most of the species sympatric in a small region must be morphologically different in order to lead different lives and so coexist successfully with the rest of the fauna. In consequence it has usually been possible to arrange museum specimens in rather well-defined groups which usually do correspond to those groups that 'breed true' in nature at any one place. As we have seen, difficulties arise over the limits between allopatric, not sympatric forms. But whenever evidence has come to light that shows morphologically distinct groups to be part of the same stock—whether individual variants, seasonal forms, castes, or life-cycle stages—no one has ever hesitated to lump them together under the same specific name. Consequently, when, as with extreme siblings, morphological criteria of taxonomic value are absent or nearly so but genetical criteria are available to delimit two or more groups that do not interbreed, it is inconsistent to cling to the morphological similarity in spite of the genetical diversity. The taxonomist has always been tracing the limits of natural intra-breeding but not interbreeding groups, and until recently has found the methods of comparative anatomy to be completely sufficient. Now, they are not, although still exceedingly valuable, and often the only methods usable.

The species is the lowest rank of groups that cannot interbreed in nature. This is what has been called the biological view of the species (as against the palaeontological, morphological, or taxonomic view). One of the best biological definitions of the species has been given by Mayr, who says:

> "A species consists of a group of populations which replace each other geographically or ecologically and of which the neighbouring ones intergrade or interbreed wherever they are in contact or which are potentially capable of doing so (with one or more of the populations) in those cases where contact is prevented by geographical or ecological barriers."

Or shorter:

> "Species are groups of actually or potentially interbreeding populations which are reproductively isolated from other such groups."

This definition deserves close examination. It is, as Mayr has emphasized, a compromise, because reproductive isolation is a practicable criterion only for sympatric, synchronically reproducing species—that is for those that do occur together in nature in a breeding condition in both space and time.

> "A population of *Drosophila melanogaster* of the year 1942 is reproductively isolated from a *Drosophila melanogaster* population of the year 1932 by a complete time barrier. Still, nobody would call them two different species. The same is true (to a slightly less absolute degree) for many geographically isolated populations. They may be reproductively isolated by a geographic barrier, but still they are not necessarily different species."

Where the reproductive isolation cannot be observed, the taxonomist must fall back on the degree of morphological difference (and behavioural if the necessary evidence is available) as an indication of probable specific difference. He must form an opinion on whether the two forms considered would interbreed if, because of hypothetical changes in the lie of the land, they were to meet in nature. The justification for this is the observation, discussed above, that good species usually do differ morphologically.

In actual practice, then, the taxonomist faced with an array of very closely related forms which replace each other geographically falls back on the morphological criterion except that he must include as subspecies under the same specific name all those forms that intergrade both morphologically and geographically. Thus in the example of the coconut lories discussed above the various forms inhabiting the northern half of New Guinea appear to intergrade where they meet, and must be called subspecies. The populations inhabiting the enormous range of islands from New Britain to the New Hebrides appear to be, morphologically indistinguishable and so similar to the eastern New Guinea forms, that they must be regarded taxonomically as a single group of subspecific rank. The western and eastern Australian forms differ rather sharply from each other, but agree in some features which mark them off from all other forms. Anyone who wishes could say that in his opinion the Australian forms are to be considered as subspecies of a separate species. He might even consider them as two separate species. No one can contradict him nor can he produce any decisive evidence for his opinion, since all the forms are wholly allopatric. Since their peculiar characters, although striking, could have a very simple genetic basis, and since the trinomial name of a subspecies indicates clearly and conveniently its relationship to allied forms, it is better to regard these (and all such forms) as subspecies.

Intergrading or interbreeding in Mayr's definition refers of course to *primary* intergradation or hybridization, that is to a smooth and gradual transition, geographically, from the characters of one form to those of the other, the intermediate populations being no more variable in their characters than are any of the extreme populations. When two forms that differ considerably in their genetical make-up come together in the wild, the resulting hybrids usually show every possible combination of characters of the two parent forms, and give the impression of a great burst of variation. The production of such hybrids by the comparatively recent meeting of two forms is called *secondary hybridization*, or occasionally simply hybridization, and is not in question here. It presents certain difficulties that are discussed later (p. 95).

The alternatives of 'geographical or ecological' replacement and barriers require explanation. Ecology might be called either 'scientific natural history' or the study of the domestic lives of animals. It considers their modes of life, the habitats they occupy, and their interrelations with their environment in the widest sense. As Elton has correctly and delightfully put it, "When an ecologist says 'there goes a badger' he should include in his thoughts some definite idea of the animal's place in the community to which it belongs, just as if he had said 'there goes the vicar.'" Geography on the other hand is the study of spatial relationships of the surface features of the earth. The meaning of geographical and ecological replacement can be best illustrated by a set of examples.

(a) The Small Cabbage White and Green-veined White butterflies are both common in most parts of Britain, but, as mentioned above, the caterpillars of the first live on cultivated plants, chiefly cabbage, while those of the second feed on wild crucifers (charlock, cuckoo-flower, hedge mustard etc.). Although there is a distinct ecological difference between the two species, their geographical ranges well-nigh coincide in Britain, and it is often possible to find their caterpillars within a few feet of each other. Adults of the two species must frequently meet in nature.

(b) In the Solomon Islands live many species of white-eye (*Zosterops*). The white-eyes are found from Africa to Samoa, and are small songbirds, rather quietly coloured, with usually a ring of white feathers round the eye. Several of the Solomons forms are closely related and obviously geographical representatives. Two species are confined to single islands. These are *Z. murphyi* on Kulambangra and *Z. stresemanni* on Malaita. The first is an olive-green bird with a yellow tinge below, a very broad white eye-ring, a brown iris, a black beak with a yellow base, and grey legs and feet. It is found in the mountains. The second has no trace of an eye-ring; the beak is grey at the base and tip, with a yellow band between, and the feet are greenish grey. According to Mayr, it is a common bird from the sea coast to the mountains. These two species show (for white-eyes) very conspicuous differences, and their altitudinal distributions are not the same. But whatever their evolutionary relationships to each other and to the other species of *Zosterops* in the Solomons, and whatever the ecological differences between them, there is no doubt that they have distinct geographical ranges, readily expressed by simple reference to the two islands on which they are found. It happens that these islands already have names, which can be found from any good atlas, and all one needs to say is—*Z. murphyi*, Kulambangra; *Z. stresemanni*, Malaita. These two species have clearly distinct geographical ranges.

(c) The ptarmigan (*Lagopus mutus*) occurs in Britain only at high altitudes in Scotland and the Hebrides. In summer it is found above the heath and heather-covered moorlands, in the regions where the dominant vegetation consists of mosses and lichens. The red grouse (*Lagopus scoticus*), on the contrary is very much a bird of the heather moors, and is more widely distributed, being found also in Wales, northern England, the Orkneys, and Ireland. Mountains being what they are, it is obvious that the ptarmigan, since it occupies the highest zone, is likely to occur in a number of colonies (in sufficiently mountainous areas) isolated by lower country suitable for the red grouse. And this is indeed what happens; it occupies islands of high ground within the territory of the other species. Only one of these two species is found (in the breeding season) in any one locality, and their ranges are mutually exclusive. But their *geographical* ranges are said to overlap, since both can be found in such a geographically recognized entity as 'Scotland,' or a particular Scottish county. All the places where the ptarmigan occurs, taken together, do not make up a single geographical entity, and to refer to them all by any accepted geographical notation (names of mountains, or latitude and longitude) would take considerable time. It is far easier to say that the red grouse is a bird of the heather moors in Scotland and the ptarmigan a bird of the alpine zone—that is, to describe their ranges by a broad geographical reference plus a broad ecological one. All species with relative distributions like those of this pair are said to overlap geographically.

From a consideration of these examples, we can draw the following conclusions.

(i) The ranges of two species, if spatially separate are called geographically separate no matter what differences in ecology there may be between the two species.

(ii) The geographical ranges of two species are said to overlap if they coincide in any area, or interdigitate extensively, or enclose islands of each other, even if at any one locality only one species can be found.

Since no two localities on the earth's surface are exactly alike in climate, geology, flora and fauna, it follows that if two species have separate geographical ranges, then almost certainly they are not living under exactly the same ecological conditions. Their ecologies are bound to differ, even if almost imperceptibly. Difference in geographical location almost always involves differences in ecology. But the converse is not always true. Two species of bird may be confined to the same patch of rain-forest, for example, but if so, it is always found that they continue to coexist by adopting different modes of life.

Further, the obvious geographical barriers that limit the distribution of many animals and plants do so only because they are tracts of ecologically unsuitable country. A great river, an arm of the sea, a desert, or a huge mountain range are both geographical and ecological barriers to certain species.

The use of 'geographical or ecological', then, in the biological definition of the species is only to cover both those forms whose distribution is more easily described ecologically ('animals of the alpine zone, or of tropical rain forest' etc.) and those that are more readily referred to geographically ('the white-eye of Kulambangra, the shrew of Islay,' etc.). As long as parasitic species are not in question (these are discussed later) there is no more to 'ecological' as against 'geographical' than that.

But we have seen that two species can actually replace each other at any single locality, while their geographical ranges may be considered to overlap widely. The ptarmigan and grouse are an example. Are such species sympatric or allopatric—that is, do they, or do they not "occur together" for the purposes of the biological definition of the species? The definition, as mentioned above, applies only to sympatric, synchronically reproducing species.

Obviously, for a beginning we must narrow down the ranges of the species concerned to the breeding ranges. If individuals of two forms never meet in nature while both are in breeding condition, then we cannot decide on their relative status. As Mayr points out, the *Drosophila melanogaster* of the year 1932 are completely isolated from those of 1942, yet they belong to the same species. He has observed large numbers of the Australian Roller (*Eurystomus orientalis pacificus*) wintering in New Guinea where there are resident subspecies of this bird. There is no interbreeding between the Australian and New Guinea forms because the former are not in breeding condition. One cannot decide from this evidence alone whether the two are actually conspecific or not. Individuals of any number of morphologically different but closely related forms may coexist in the same locality and meet frequently, without providing any evidence about their relative taxonomic status.

Secondly, we must be sure that individuals of the forms in question are actually able to meet in such a way that interbreeding can take place. Many species have very strong habitat-preferences. If the dispersal phase in the life-history of a species is a larval stage, as so often in marine animals, habitat-selection may be mainly passive—those that meet the right conditions survive, the others die. But when the dispersal phase is an adult or subadult stage, there may be active searching. Migrant birds returning to their breeding areas provide an excellent example. If the areas suitable to one form are always separated by a thin belt of unsuitable country from those utilized by another, the two forms, although occupying ranges that coincide geographically, may never actually meet in breeding condition. I am indebted to Mr. R. E. Moreau for a striking example in non-migrants. Certain forms of white-eye (*Zosterops*) are confined to isolated patches of mountain forest at high altitudes in East Africa. A different form is found in the lowlands and floods around the highland ones which occupy islands in its range. But although the lowland form sometimes extends far up the lower slopes of the mountains, it seems that there is always a gap, often remarkably narrow, between it and any of the mountain forms. The latter seem to be extraordinarily faithful to their type of habitat, and often differ conspicuously in plumage characters from the lowland form. But because there is no actual meeting, one cannot decide on the relative status of these forms.

It might be thought that two forms could be said to coexist in the same area if there were no spatial barriers between them. The populations of monkeys, for example on the two banks of a large river might not be completely separated if they could swim. Jackdaws nest abundantly in low cliffs around the Orkney Islands, and hooded crows are common on the moors. The two species are certainly not separated by any spatial barrier that they cannot overcome in a few minutes. But one must be very careful in deciding what can be overcome. The physical capacity for locomotion is one thing, its employment is another. The distance between some of the mountain forms of African *Zosterops* and the lowland one is quite trivial yet it is not certain that it is actually traversed. Mayr points to the *Zosterops* of the Solomon Islands as another good example. On several islands of the Central Solomons live forms which are so distinct that they are given specific status; in this small geographical area is a superspecies with several species and subspecies. Some of the islands are separated by only a few miles, and these birds could easily cross such gaps in a few minutes. Yet their distinctiveness on each island is such that one is forced to the conclusion that they seldom, if ever, do.

Many such examples could be given; in them it is not the spatial barrier but the habitat-preference of the birds that is the restricting factor. Geographical separation is brought about by an intrinsic barrier. In consequence, one cannot say that forms prevented from interbreeding only by intrinsic factors can be regarded as sympatric, since the forms of *Zosterops* in the Central

Solomon Islands mentioned just above are clearly allopatric. Yet it is obvious that many animals that are found in breeding condition and in close spatial proximity must be prevented from interbreeding successfully by intrinsic factors; they are hindered only by their own behavioural patterns, or genetical incompatibility. The slugs *Arion hortensis* and *A. circumscriptus*, the snails *Cepaea nemoralis* and *C. hortensis*, the Large and Small Cabbage White butterflies, the song thrush and mistle thrush, the fresh water mussels *Unio pictorum* and *Unio tumidus*, the flies *Drosophila obscura* and *Drosophila subobscura*, the earthworms *Lumbricus terrestris* and *L. rubellus* are just a very few pairs of very common British species that must be prevented from interbreeding primarily by intrinsic factors.

The difference between such pairs of species as these, and the various forms of the *Zosterops rendovae* superspecies in the Central Solomons is that in the former, individuals of the two species do actually meet in nature in such a way that *attempts at pairing* could take place, whereas it seems that in the latter they never do. This gives us the means of distinguishing between allopatric and sympatric species. Sympatric species are those in which some individuals from each in breeding condition come into such spatial proximity that fertilization is possible, allopatric species are those in which this never happens.

These definitions help us to resolve an ambiguity which arises over geographical replacement and allopatry. If two closely related species are completely separate spatially, they show geographical replacement and are also truly allopatric from the genetical point of view. If their ranges meet without overlapping, but are *contiguous* along a common boundary, then they are still geographically replacing, but if pair-formation along the boundary (the edge of a rain-forest, for example) is possible, they are genetically sympatric and their relative status is determinable. Geographical replacement (in the sense that only one of two species is found in any locality) and genetical allopatry are not exactly the same thing.

It seems advisable, therefore, to omit from the biological definition of the species any reference to ecological or geographical replacement, and to refer explicitly to genetical sympatry and allopatry. A suitable modified definition might run as follows:

At any one locality, numbers of individual organisms may be

found which are genetically sympatric, that is, they coexist in such a way that inter-breeding between all of them is possible, yet which fall into well-defined groups within which but not between which interbreeding produces normal offspring. Such groups are members of distinct species. When such groups in different localities are found to be interconnected by spatially intermediate populations such that *secondary* hybridization is not taking place, all such groups are referred to the same species. Unconnected (allopatric) groups may also be referred to the same species if it is considered that their similarity warrants this step.

Nothing whatever is said in the biological definition of the species about the behaviour of forms brought together artificially. By 'artificially' is meant solely because of human interference to that end. The production of a zoological garden in London and the imprisonment of several species of monkey in one cage in it is artificial; the joyful exploitation of it by hordes of London sparrows is not. Similarly, the production of hybrid flowering plants by careful cross-pollination with a small paintbrush is artificial. The existence of huge numbers of hybrid hawthorn trees in North America has been ascribed to man's interference with the landscape by lumbering, and cultivating, since many forms now meet (and interbreed freely) that formerly would not have met at all. If this is correct, the hybrid hawthorns cannot be regarded as artificially produced.

The reason for this absence of any reference to artificial crossings is that such crossings tell us almost nothing. Many birds are known that behave as good species in nature, but can be made to hybridize freely in aviaries. And there are many species that refuse to breed at all in captivity. It is necessary to make a clear distinction between the capacity to produce fertile hybrids and the exercise of this capacity in nature. Mayr refers to the former as *fertility* between any two species, and to the latter as *crossability*. Two species may be completely interfertile in artificial conditions but show no crossability at all in nature. But of course it is true that if two forms both live and breed successfully in cages but either refuse to mate with each other, or mate but produce no offspring, or produce sterile or weakly offspring, then one can conclude that they are specifically distinct. It is not true that if they interbreed successfully, they are by that very fact not good species, because they may show no crossability in the wild.

The species has been described above (p. 52) as the lowest rank of groups that do not interbreed in nature. Nevertheless it is true that certain forms are considered good species although they hybridize regularly. In fact, one can construct an almost continuous series beginning with closely related species that never hybridize and ending with certain plant species which seem to hybridize in some regions to such an extent that it is rare to find specimens which are not hybrid. There is no consensus of opinion on when hybridization is too small to be of importance in determining specific limits and when it is so high that the hybridizing forms must be regarded as having combined to form a new species.

In animals but not in plants, hybrids are usually very rare, or confined to certain rather well-defined 'hybrid zones'. Sporadic hybridization is known in many species of birds. In some it is clearly associated with rather abnormal conditions. Lodge has stated that the capercaillie (*Tetrao urogallus*) in Scotland sometimes make sudden local movements. The females go first, the males not usually following until the next year. Before the males join them the females not infrequently mate with males of the black grouse (*Lyrurus tetrix*) and produce hybrids, many specimens of which are now known. But these hybrids seem to remain uncommon, and it does not appear that there is any very effective gene-flow (i.e. interchange of hereditary factors) between the two species. A rather similar example of hybridization has been reported in two American game birds. An island in Lake Huron has recently been colonized by both the sharp-tailed grouse, *Pedioecetes phasianellus*, and the prairie chicken, *Tympanuchus cupido*, the latter being much the more abundant. There is considerable hybridization between these two normally distinct species, presumably because the *Pedioecetes* have difficulty in finding mates of their own species.

One example of a hybrid zone has already been given, namely that in the great-tits of Khorassan (p. 56). A much larger and better-known one occurs between the carrion crow (*Corvus corone*) and the hoodie crow (*Corvus cornix*) where their ranges meet in Europe and northern Asia. In Europe, the carrion crow is found in England and south Scotland (only rarely in

North Scotland) and in Spain, France, the Netherlands and Denmark. The hoodie crow, which is very like the carrion crow except that its body is grey, only the head, upper breast, wings and tail being black, is resident in Ireland, the Isle of Man, north Scotland, and Europe east of the carrion crow's range. In a narrow belt in Scotland and again in a long sinuous zone extending from the north German coast to north Italy, hybrids between these two forms are common; in fact Mayr states that a careful examination in certain localities showed that no purebred crows were present. "Even birds that seemed in the field to be either *cornix* or *corone* showed their hybrid nature on closer examination in the laboratory." The width of the hybrid zone is only from seventy-five to one hundred kilometres, and it appears to be remarkably stable. One would expect that if there were no barriers, hybridization would spread all over Europe, but there is no indication that this is happening.

The hybrid zone is clearly recognizable because of the great outburst of variation within it. To the south and west are found in the breeding season very uniform populations clearly belonging to *C. corone*, to the north and east equally uniform ones obviously assignable to *C. cornix*. In the zone every possible intergradation is found between these two forms, so readily distinguishable by their colour-pattern. Still, although the intergradation is secondary, it is none the less intergradation, and some authors have suggested that the two forms should be regarded as subspecies of a single species. But many have felt that since the hybridization is not primary (p. 58) and since there is no sign that it is producing anything more than a very restricted effect, it is better to regard them as distinct species, which they are over the greater part of their ranges. Obviously, their rank is a matter of opinion since they are almost entirely allopatric.

To sum up, the distinctiveness of two species is recognized primarily by the fact that they hybridize very rarely, if at all, when they meet in the wild. Usually since any two species are always specialized for leading rather different lives, they differ morphologically as well as in behaviour, physiology, ecology and genetics. Without genetical isolation, they would merge into a single form. The biological definition of the species emphasized its genetical isolation, which allows it to display its own peculiarities. In some closely related species morphological differences may be slight or even practically absent. There is every possible intergradation in degree of morphological difference between such sibling species and more 'normal' ones. Even though specimens of some extreme siblings may be inseparable if only preserved adults are available, that is no reason for denying them the status of species. Morphological distinctiveness is only a general, not an infallible guide in delimiting species.

CHAPTER VII

OTHER SORTS OF SPECIES

THE introduction of the biological definition of the species is a considerable advance, not only because it takes full account of geographical variation and of crossability as well as of morphological resemblances, but because it cannot be applied to certain animals. This is not paradox for the sake of paradox; formerly at least three different meanings were confused under the term species, and the development of one of them (the biological species) has clarified the others.

AGAMOSPECIES

The biological definition of the species can be applied only to sexually reproducing animals (and plants). But many animals never reproduce sexually at all. Some plants, especially garden varieties with sterile 'double' flowers, produce no seed and can be propagated only by taking cuttings, by layering, or by some other device for producing *vegetative reproduction*. In Great Britain the Canadian Pondweed (*Elodea canadensis*) is an introduced water-weed which has spread very widely. The male and female flowers are borne on different plants, and it appears that only females have been found in Great Britain. The whole of its spread in this country is due to vegetative reproduction. Many animals such as *Hydra*, some flatworms, and probably all sponges, can reproduce by budding; perhaps some reproduce only vegetatively. A more usual process which does not involve true sexual reproduction in animals is *parthenogenesis* (virgin birth). In this process, eggs are produced by females, as for sexual reproduction, but they develop without fertilization by a spermatozoon. Sometimes parthenogenesis and normal sexual reproduction with fusion of eggs and spermatozoa are regularly found in the same species. In many waterfleas and greenfly parthenogenesis is used during favourable conditions (especially in the summer). The young, all females, are either brought forth alive, or are hatched from soft-shelled 'summer eggs'. On the approach of less suitable conditions some males are produced which then fertilize the females in the normal way. These females lay resistant tough-shelled 'winter eggs' which can last through periods of adversity. Such forms as these can be brought under the biological definition because sexual reproduction does occur at some stage and the criterion of actual or potential interbreeding can be used.

But in many diverse animals, parthenogenesis has become obligatory; either males are never found, or in hermaphrodite animals, the male set of genitalia does not function. Such a state of affairs seems strange to most people because the animals they know best—the vertebrates—indulge in neither vegetative reproduction, nor frequent nor obligatory parthenogenesis. Some plants use all the appurtenances of sexual reproduction without actual fusion of male and female gametes. Many species of dandelion, for example, are *apomictic*; they have fully developed florets which produce both ovules and pollen, and for development to occur, pollen must fall on to the stigma as in normal production. But the pollen tube does not convey the male gametes to the female, and although pollination is necessary for development, there is no actual fusion of gametes. Such organisms as these apomicts and obligatorily parthenogenetic species cannot be brought under the biological definition because no two individuals ever contribute to the hereditary endowment of a third; every individual has only one parent, not two.

At first sight, this may seem no great difficulty. One can still group the forms by their morphological similarities into species and subspecies. This is of course true. It is one of the great advantages of the morphological species that it is purely morphological and can be applied to any organism irrespective of its mode of reproduction. But the morphological species is merely the lowest taxonomic rank (Chapter V). Well-marked forms might be considered varieties, subspecies, or even full species if they do not actually intergrade with their closest relatives. Exactly the same considerations hold here.

One of the commonest of all earthworms in the Northern Hemisphere (and elsewhere by introduction) is the amphibious species *Eiseniella tetraedra*. It is about an inch long, and yellowish brown. In the posterior half of the body the four pairs of bristles in each segment, by means of which it can progress, are about equidistant from each other and are mounted on slight ridges, so that in cross-section, it is square posteriorly, rounded anteriorly. It is very abundant on the banks of rivers, streams and even tiny brooklets where grass or other vegetation hangs into the water, and in shallow, swift-running or otherwise well-oxygenated water it may be totally aquatic. It is in all respects a normal member of the *Lumbricidae*, possessing a pair of ovaries, with funnels, egg-sacs, oviducts and female pores, and two pairs of testes, with large accessory sacs, funnels, male ducts, and male pores. As usual, there is in sexually mature individuals a conspicuous glandular band, the clitellum, which secretes the cocoon. The clitellum has special glandular areas, exactly as in its close relatives *Allolobophora caliginosa* and *A. chlorotica* described above (p. 75). Like them it appears to be a cross-fertilizing hermaphrodite and possesses special sacs, the spermathecae, to receive sperms during mating. (In mating two earthworms come together and each transmits its own sperm to the other individual's spermathecae. At convenient times after they have parted each worm forms cocoons into which it puts its own eggs, and sperm received from the other individual. Fusion of the gametes (the eggs and sperm) may take place months after mating).

Eiseniella tetraedra is readily recognized by the following additional characters. Male pores on segment 13 with large swollen lips; clitellum from segment 22 or 23 to 26 or 27; the glandular ridges on segments 23 to 25 or 26. The spermathecae open near to the mid-dorsal line, in the furrows between segments 9 and 10, and 10 and 11, not laterally as in the species of *Allolobophora*. The species is a well-marked one, the genus is closely allied to the genus *Eisenia*, but fairly clearly separable. Most of the well-known Lumbricid earthworms reproduce entirely sexually, although occasionally eggs may develop parthenogenetically. There is nothing in *Eiseniella tetraedra* that marks it as in any way abnormal. Yet it is totally parthenogenetic. Both it and several other species in other genera of earthworms, which on the criteria of general morphology and of distinctness and absence of intermediates appeared to be as 'good' species as any are now known to be incapable of sexual reproduction. Testes and all the accessory male organs are present, and the process of producing spermatozoa actually begins. But it is never completed. The spermatozoa abort; no fully formed ones are found in their normal place in the sacs accessory to the testes, and none are to be found, transferred from a different individual, in the spermathecae. It is possible, from the presence of the complete set of male genitalia, that parthenogenesis is comparatively recent in such forms as these, too recent for any loss of unwanted organs to have occurred. In some species the spermathecae may be occasionally or permanently absent; presumably with the supervention of parthenogenesis they are no longer required and can be allowed to disappear with impunity.

In *Eiseniella tetraedra* there is considerable variation, and several clear-cut forms are known, some of which have at different times been named and ranked as a species, a subspecies, a *forma*, a *mutatio*, or various other infraspecific categories. The table gives the characters of the principal ones.

Some varieties of *Eiseniella tetraedra*

	Male pores	Clitellum	Glandular ridges or patches
Eiseniella tetraedra			
forma *typica*	13	22 or 23–26 or 27	23 or ½ 23–25 or 26
f. *hercynia*	15	,,	,,
f. *tetragonura*	13	18–22	19–21
f. *macrura*	13	15–22	20–21

What rank should be assigned to these various forms? No definite answer can be given. All occur without intermediates side by side with the 'typical' form (that is, the commonest, which was named first), and could therefore be considered separate species. If they interbred freely they would certainly be only well-marked varieties; if they could interbreed but did not do so, they would certainly be good species. But there is not even the possibility of interbreeding, so we cannot say. The offspring of each individual are genetically nearly identical with their parent, except when mutations have occurred, and probably resemble it and each other far more closely than any other individuals. Almost certainly the various forms shown in the table, now they have arisen, 'breed true'. Genetically, the progeny, for any number of generations, of a single individual are all connected by their descent. Suppose that among them one individual appears which differs visibly because of a mutation. It is equally connected by descent from the common ancestor, but differs morphologically from the rest and again will 'breed true'. Is it to be regarded as a separate species? Again, we cannot tell since there are no criteria for deciding what is a species and what is not.

Obviously, mere community of descent will not help in the slightest. There is very good evidence that birds are descended from one group of extinct reptiles, mammals from another, and reptiles themselves from certain very primitive amphibians. Living mammals, birds and reptiles are all descended from a common stock but they are not therefore all the same species. If the sequence of fossils were more complete, one might see on a single line of descent such profound although gradual changes that the forms at the two ends must be in separate classes. Community of descent as such tells us nothing about changes during the process of descent.

The biological definition of the species is not subject to this difficulty because it uses as its criterion not genetical continuity in the past (which in theory at least can be pushed back in time until it involves all animals living and extinct) but the potentiality or actuality of combining to make joint contributions to the genetical make-up of the next generation in the future. One therefore assesses the status of different forms by their actual or presumed genetical relationships in the present, or rather over a short stretch of time which includes a breeding season of either or both the forms under investigation. If their breeding seasons never coincide, or coincide without interbreeding taking place, then these two forms are known to be different biospecies. This short period can be called the *time-quantum of the biospecies*. If the two forms cannot be observed in the wild during one of these quanta, then they can be given specific status only on morphological evidence. The usefulness of the definition of biospecies in giving a more than merely positional value to the species lies in the restriction of its application to the time-quantum, during which any biospecies can be observed to be genetically isolated from all others.

It seems advisable to recognize clearly the limits of the biological species, and it is convenient to distinguish those forms to which it cannot apply because they have no true sexual reproduction as *agamospecies*. Species-criteria for such forms are the same as those given above in Chapter V for the morphological species.

Some agamospecies are so recent that they can be distinguished from sexually reproducing forms only with difficulty. In several species of small moths, some stocks are parthenogenetic, others not; apart from this difference, they may be almost indistinguishable. Such forms present considerable nomenclatural difficulty. Their evolutionary potentialities are very different. The whole point of sexual reproduction, the reason for its employment by the great majority of living things, is that it provides a means for producing constant genetic variation without causing disintegration of the exceedingly intricate and complex system of processes that is the living organism. The raw material of variation is provided by mutations, that is, by apparently spontaneous changes in the hereditary material. But the mutations are sudden changes which seem to have no reference whatever to the structure or needs of the individual in which they occur. In that sense, they may be called random changes. The chances of a large mutation being beneficial are therefore very much less than the chances of hitting a ship's chronometer with an axe and improving its working. Small mutations may be tolerated, especially since living organisms, unlike machines, have considerable powers of compensatory change, which can lessen the bad effects of any sudden shock. It seems likely from the investigations of geneticists that no individuals except those propagated vegetatively (including identical twins) have exactly the same genetical set-up or *genotype*. New mutations (if they can persist at all and do not make their possessor incapable of breeding) can be transmitted to the progeny, whose genotypes will be only a few of the enormous number of possible combinations of different alternative characters from their two parents. Thus the new mutant will be constantly transmitted into different genotypes. Moreover a genotype is not merely the simple sum of all the hereditary factors, or *genes*, in it; there is very considerable interaction, sometimes, between the products of the genes. Consequently a new mutation, constantly introduced by sexual reproduction into a variety of genotypes made up of different selections from the genotypes of both parents, can be tried out, as it were, in combination with all sorts of other genes. The chances of its proving useful eventually are far higher than if it had to stand or fall by the effect it produced in the individual in which it happened to arise.

Mutations are the raw material of genetic variation, which is mainly produced by their constant shuffling and assortment by means of sexual reproduction. The mutation rate can thus be kept quite low, so that the chances of a single individual succumbing from the shock of several large mutations are exceedingly slight, while genetic variation (by continual recombination of mutations established in a species) can be considerable. Every species is subject to all sorts of pressures from natural selection. If circumstances change for any reason, such as a major change in world climate, or an invasion of a district by a potential competitor, a species with high variability has a good chance of persisting, since many individuals may prove to be fairly well adapted to the new situation. An agamospecies which by comparison with sexual species has very little variability has lost the capacity to adapt itself to changes. It may be extremely successful for a time, more so than many sexually reproducing forms, but if conditions change, it is much less likely than they to be able to survive.

Consequently, to a very large extent, agamospecies can be regarded as off the main stream of evolution. Their temporary success may be great but it will be only temporary. The distinction between sexually reproducing species and agamospecies is therefore very important from the evolutionary point of view. But it is true that when parthenogenesis first becomes established in a particular individual of a population, its genetic basis may be very simple indeed, and a reversion to sexual reproduction may be very easy. Indeed, as mentioned above, a regular alternation between parthenogenesis and sexual reproduction is normal for some species in very diverse groups of animals. In consequence, one often feels that when a particular strain of a moth, say, is parthenogenetic but morphologically indistinguishable from other strains, to separate off this one strain as a distinct species with a formal name of its own is a rather hasty procedure. Morphologically, as in the Lumbricid earthworms, agamospecies and sexually reproducing species may be found intermingled in such a way that it is evident that parthenogenesis has arisen readily and frequently. In some species, such as the freshwater worm *Lumbriculus variegatus*, a prolonged examination might reveal, even in this very abundant species, only vegetative reproduction by fragmentation, which is exceedingly common. Yet very rarely a number of sexually mature specimens are found together in one locality, which are producing eggs in cocoons apparently after true sexual reproduction. This species is not, after all, an agamospecies, although it seems to be very near to that status. As with sibling and 'good' morphological species, every degree of intermediacy is known between agamospecies and those, such as man, which can reproduce only bisexually.

The biological species concept, then, cannot be applied at all to forms without true sexual reproduction. Yet these forms have clearly arisen very many times, and must be classified next to their nearest known relatives. Moreover, many forms known as yet only as morphological species may well be agamospecies. The morphological definition can be applied as well to them as it can to forms with true sexual reproduction, since it takes no account of modes of reproduction, and they cannot as yet be recognized as agamospecies. There is no difficulty in fitting agamospecies into the natural hierarchy based on morphological comparisons —on the contrary, the difficulty is to pick them out. Practically, one can merely recognize that in the category of the known morphological species may be many agamospecies, and that when a form is known to have abandoned sexual reproduction completely, it can be associated with its sexual relatives only by means of morphological similarity, and can be assigned no certain status in the natural hierarchy.

Dobzhansky has gone so far as to say that "the species as a category which is more fixed and therefore less arbitrary than the rest is lacking in asexual and obligatory self-fertilizing organisms. All the criteria of species distinction utterly break down in such forms. The binomial system of nomenclature, which is applied universally to all living things has forced systematists to describe 'species' in the sexual as well as in the asexual organisms. Two centuries have rooted this habit so firmly that a radical reform is beyond practical possibility. Nevertheless, systematists themselves have come to the conclusion that sexual species and asexual ones or 'agamospecies' must be distinguished. . . . All that is saved by this method is the word 'species'." One must agree to some of this, adding only that the purely morphological species is no less unfixed a rank than the agamospecies, yet as a cataloguing device it is probably indispensable: and that the word 'species' as a rank name in the hierarchy can be applied to agamospecies and bisexually reproducing ones as well on theoretical grounds as by force of habit. The old idea of the species is now subdivided: one can either restrict the name to one meaning (presumably the biological species) or use it with appropriate prefixes for all. Either course is good, but the second is by far the most convenient. Binomial names standardized and protected from homonymy by the International Rules, are a very convenient means of reference, and when different strains of agamospecies have important differences for the plant pathologist or medical man, the shortness of the binomial has often caused them to be used in preference to trinomials. It is greatly to be desired that some system of simple prefixes should be agreed upon, so that the status of a binomial should be clearly indicated, and that morphological, biological and agamospecies should be recognized as such.

PALAEOSPECIES

It has already been pointed out that according to the theory of evolution any two groups of animals alive today had a connecting common ancestor at some time, more or less remote, in the past. In consequence, all the discrete natural groups recognizable today can be represented as the terminations at present of a number of branches of a single vast evolutionary 'tree'. A rough sketch of that branch which ends in the modern Carnivora is given in fig. II (p. 39).

The discreteness of the natural groups of animals actually recognized depends therefore on two facts, namely, that one does not find every possible intermediate between any two species of animal existing in any one short period of time, and that the fossil record, except in a very few groups, is extremely imperfect. To take as example that period that we know best zoologically, the present day, we find, as for example with the gibbons, great apes and man (fig. I, p. 36) that individual animals do tend to fall into numbers of discrete groups, and that these groups usually cannot be interconnected by means of fossils, since too few are known. The evolutionary history of man, for example, must be reconstructed from exceedingly incomplete evidence.

The imperfections of the fossil record are very useful. Because of them, the known fossils of most groups also fall into rather discrete assemblages, and the hierarchical classification, devised originally for living forms, could therefore be applied without modification to fossils. As fossils were discovered they were simply grouped with their known relatives and ranked according to their likeness to them. Such an extraordinary animal as the giant sloth *Megatherium*, for example, is nevertheless obviously a mammal, and related to the present-day anteaters, sloths and armadillos of South America. It is at present placed with many other extinct genera in the same order and suborder as the living South American anteaters. But when good series are available, forms that seem to be good species at any one time may become indefinable since they are successive stages in a single evolutionary line and intergrade smoothly with each other. It is convenient to refer to such forms as *palaeospecies*.

It is interesting to reflect on what system of classification might have been adopted if for some reason good series of fossils were so well known to mankind that living animals were recognized from earliest times as the present terminators of evolutionary series; or alternatively if the process of evolution in large groups of economically very important and therefore carefully studied insects were to be visible over a period of some hundreds of years. The static monotypic morphological species (or other rank) could never have assumed any great importance. The 'units' of classification must have been the separate evolutionary series, often called *gentes* (singular, *gens*) or *phyla* (not to be confused with *phyla* used as a rank name in the natural hierarchy). And since several of these same 'units' could obviously arise by the splitting of a single one, the reference system used must make provision for indicating this fact, and dealing adequately with transitional forms. The system used at present, developed for reference only to living animals over a period of time so short that evolutionary change is practically imperceptible, does neither of these things.

In a discussion of the species concept as applied to fossils, there are two quite separate points to be raised. One is the relation of the species concept to time. The other, of less general importance, concerns the use of binomial names by palaeontologists, and can be disposed of first.

It must be remembered that many palaeontologists are primarily stratigraphers, to whom fossils are of the greatest use in recognizing and correlating particular strata. A close study of the contained fossils is often essential, always desirable, in arranging the rocks of a district in their time-sequence and determining their detailed correspondence with those of other districts. For these purposes, a particular sort of fossil which is abundant in very many different strata is not of much use, while another which is found only in a very narrow band of rock can be used for much more detailed determination. An isolated outcrop of rock containing only fossils abundant throughout the Jurassic period, for example, could be merely assigned to the Jurassic on its fossils; but if some of them were known to be found elsewhere only in one series of beds in the Lower Middle Jurassic, there would be the strongest presumption that this isolated outcrop was contemporaneous with these beds. Consequently, in stratigraphical palaeontology, the greatest emphasis is laid on those fossils with a very restricted distribution in time, since these permit the most precise stratigraphical identifications and correlations; and it is very convenient for such fossils to have brief and officially recognized names.

It has happened rather frequently in the past that, as with living animals and plants, a 'species' has been named from only one or two specimens, and has since been discovered to be based only on extreme individual variants, connected with another 'species' by all grades of intermediate forms. In the study of living things (neontology) such a discovery means that the younger of the two specific names must be treated as a synonym of the older, because only one species must be recognized. But in palaeontology this procedure has not always been followed. In successive strata apparently good species can be seen to be slowly transformed into others, so that morphologically distinct forms which are successive stages of the same gens are interconnected by intermediates, and the breaking of a gens into species is an arbitrary procedure. But also, some extreme variants may be common in a fossil population from one stratum but be very rare in others. If a few specimens corresponding to the morphological description of the extreme form are obtained from a stratum, they suffice to identify it, no matter whether some specimens intermediate between them and another form in the same stratum can be found or not. The stratigraphical value of a particular (morphological) form depends only on the distribution of that form, irrespective of whether it can be connected to others by a series of morphologically intermediate specimens or not. The binomial nomenclature being convenient, many forms, although known to be only individual variants that could not possibly be called species in neontology, are still listed as species in palaeontology. This practice has been condemned by several modern palaeontologists, who point out that intermediates between successive 'good species' have a completely different biological significance from those between contemporaneous ones. The first show a gradual evolution within gentes of species distinct from one another at any one time, while the second show merely that specific limits have been wrongly drawn. Others point out that as stratigraphers they are interested not in the biological significance of species, but in a suitable classification and reference system for fossils. They see no reason therefore, for upsetting a system of classification and nomenclature which has worked very well for many years.

If stratigraphical and biological research were two independent activities, there would be no confusion; but unfortunately, species can be named by any palaeontologist, whatever his particular interest, and must be taken into account by any worker who wishes to study a fossil group. The binomial names given for convenient reference to certain extreme variants of sea urchins, say, by a stratigraphical palaeontologist may be duly listed and suppressed as synonyms by any biological palaeontologist who wishes to prepare an account of the course of evolution in these animals. The stratigrapher views with irritation the suppression of his names and declines to use the cumbrous trinomials or quadrinomials which are biologically more correct; and the biologist, faced with the necessity of getting into their biologically correct status a vast number of binomial names, finds his task made harder by the activities of the stratigrapher. As with the agamospecies, if it is really necessary to cling to the use of binomials merely because they are brief, a prefix to indicate the biological status of forms which are 'species' only because they are given specific names, would simplify the situation enormously. In the meanwhile one can only bear in mind that a binomial applied to a fossil may refer only to extreme variants of a single population.

The relation of the species-concept to time is a point of very much greater importance than that just discussed. The species as it was originally employed was monotypic, and its transformation into the polytypic species when the facts of geographical variation came to be appreciated has already been discussed (p. 53). The fact that so many fossil species are known from very isolated and often fragmentary specimens meant at first that the monotypic species-concept could be applied to them without any trouble. But just as increase in knowledge about living species brought to light whole series of geographical variants, so a corresponding increase in knowledge of fossils produced series of chronological variants.

As already mentioned, species which are 'good' at the present time consist in effect of populations of individuals which at least potentially are genetically connected, since any one is capable of interbreeding successfully with any other. The gene-pool from which the genotypes of the next generation can be drawn comprises all the individuals within the species in the present generation, and no others. While good species can of course react upon each other as competitors, predator and prey, parasite and host, or in many other ways, there is no direct genetical connexion between them. The next generation of Green-veined White butterflies in the world will be the progeny of Green-veined Whites only, and will receive no genetic contributions from Large Cabbage Whites, for example. The same generalization will hold true in the next successive generation, and the next, and so on into the future, and must have been true for preceding generations up to the time when these species diverged from the ancestral stock and became 'good' species. But there is no reason to believe that they have kept exactly the same characteristics ever since they first became distinct; on the contrary, the abundant direct evidence of 'descent with modification', as Darwin expressed it, in fossil series, plus a vast body of indirect evidence leads us to expect that transformation of species in time, although very slow and gradual, is well-nigh ubiquitous. At a period half-way between the present day and the time when these two species of butterflies first became distinct, they may have had characters so different from those obtaining at the present day that, could we but know them, we should be forced to call the two forms by different specific names from those now in use. Being as distinct from their descendants as these are from each other and from their closest relatives, these ancestral forms, to be classified in a consistent natural system, must be given specific rank. But with the passage of time, they change continuously and are gradually transformed into the two modern species, without any sudden discontinuity which could be used as a specific boundary.

In fact, as soon as we begin to consider the relationships between good species and time, it becomes evident that each represents only the terminator at any one time of a gens, or as Simpson has described it, " . . . a phyletic lineage (ancestral-descendent sequence of interbreeding populations) evolving independently of others, with its own separate and unitary evolutionary role and tendencies. . . ." The various intergrading forms which compose it can be called *transients*. 'Evolving independently' in Simpson's definition refers of course to genetical independence only, as he emphasized; he refers to the concept of the species as a part of a gens as 'genetical-evolutionary'.

The recognition of gentes or phyletic lineages as separate evolutionary 'units' is of fundamental importance. As Simpson points out, all the various sorts of species recognized so far, except agamospecies, are revealed as different attempts to delimit gentes. All the evidence available is graded in importance for this end and criteria which are normally accepted may be abandoned if it can be shown that they are useless for this purpose. For example, morphospecies are in general given specific status because different species 'good' on every other criterion do normally show different morphological peculiarities. But if the genetical criterion indicates that two genetically separate populations are present in a morphologically single one, then these must be given specific rank even though there may be only the faintest morphological differences between them. *Drosophila pseudoobscura* and *D. persimilis* are correctly ranked.

But equally, two forms may be regarded as good species even though there is continual hybridization between them provided that it is either very rare or highly localized. Then, in spite of the hybridization, the two forms do not lose their identity; gene-flow is only local and its effects are confined to a small proportion of the total populations. It is true that, as shown above (p. 95) there is every intermediate between good species that apparently never hybridize, and (in plants especially) forms that were apparently good species but on meeting appear to interfuse completely. Recognition of the species as a time-section of a gens does not allow one to introduce a break into this series, but it does give the reason why a limited amount of hybridization is not generally regarded as direct evidence for the infraspecific status of the hybridizing forms.

Similarly the emphasis laid on intersterility as a criterion of specific distinctiveness, is seen in this light to be justified. If two forms breed freely in captivity, so that captivity as such appears not to affect them in any relevant way, then a demonstration that under these conditions they cannot produce hybrid offspring, or can produce only sterile ones makes it certain, as Simpson points out, that they are now in separate gentes and can be expected to remain separate. But it does not of course follow that forms interfertile in the laboratory do interbreed in the field. Forms which on Mayr's definition (p. 94) do not show crossability in the wild but do show interfertility in the laboratory are also good species, parts of separate gentes; the prediction that they will remain separate may be only slightly less certain than that for intersterile forms.

But the recognition of the importance of the gens endows the species (except agamospecies) with special evolutionary and taxonomic significance only at the expense of denying it any except arbitrary criteria in time. As long as the fossil record of any group remains fragmentary, the species recognized will be distinct purely from lack of evidence. But as soon as any gens showing considerable alteration in morphology with time can be constructed from the available evidence, then the problem of breaking it up into species becomes acute. When good evidence can be found that a single gens has forked, the situation is worse still. The gens itself can no longer be considered even in practice as a discrete unit, since it intergrades with two others. Suppose that a series of successive forms, *Aa*, *Ab*, *Ac*, *Ad* are known which show every sign of being the transients of a single gens, separable only because of gaps in the fossil sequence, and that in the strata immediately above those containing *Ad* are found two gentes, *Ao*, *Ap*, *Aq* and *Ax*, *Ay*, *Az*. If *Ao* is far more like *Ad* than *Ax* is, it might be the most convenient course to separate *Ax*, *Ay*, *Az* as a side-line and consider the line *Aa-Aq* as a single gens. But if *Ad*, *Ao* and *Ax* are all very similar this course may be impracticable. There is then no more reason in practice than there ever was in theory to consider *Aa-Aq* as a single gens with *Ax-Az* as a side-branch, than to consider *Aa-Az* as 'the' gens, and *Ao-Aq* as 'the' side-branch. On all evolutionary and genetical grounds, both limbs of the fork are equally continuations of the stem. If it were known that one limb quickly became extinct while the other gave rise to many more branches, one would be inclined to designate the second as the main limb. But it might well happen that the earliest species known in it differed more from the latest in the stem than did the earliest in the 'mere' side-branch, in which case the stem and side-branch must be taken as one natural group, the continuation of the stem another.

As long as the various forms known are called species there may be no greater difficulties than those described above. But anyone may claim at any time that several successive transients with binomial names really did not differ so greatly as to deserve specific rank. They should be considered only as chronological subspecies. The limits of both subspecies and species within a gens are equally arbitrary, since there is no reason to make a break in a continuous series at any one point rather than at any other. In consequence one might easily arrive at the situation in which by morphological standards the earliest transients in each of the two limbs of a fork must be classified with their immediate ancestor, the highest transient of the stem, as subspecies of one species, although it is perfectly clear that the two limbs of the fork are biospecifically separate. Then *Aab* and *Aac* would be equally subspecifically separable descendants of *Aaa*, yet behave as good species towards each other. An exactly analogous situation arises in neontology in the nomenclature of ring-species.

The great tits have already been described (p. 55). In them there is a complete morphological series from *Parus major major* to *P.m. minor*, yet these two coexist without interbreeding in one district (p. 57). In this example one can avoid this difficulty by recognizing the Khorassan birds as a hybrid population between two separate species, *P. major* and *P. minor* (p. 56). In other forms, with no obvious secondary hybrids, even this procedure cannot be applied. The whole system of classification and nomenclature is devised for applying discrete symbols of reference to discrete groups of animals, and it fails completely when one tries to apply it to groups discrete in one place or period but with continuous intergradation in others. While in practice one can often shirk the issue since only a few fragments of the continuous series are known, in theory it remains insoluble.

Uniformity is desirable even in arbitrary procedures. Most ornithologists agree that two populations, even if they obviously differ in average characters, should not be given subspecific names unless in the available samples, 75 per cent of the specimens in one group can be clearly distinguished from 100 per cent of comparable specimens (of the same sex or age) in the other group. Similarly some palaeontologists have adopted the following criteria. If two populations of fossils from the same gens show statistically significant difference in the mean value of some character, but overlap in their range of variation, they can be considered as different subspecies of one species. If in any well-defined character they show no overlap, so that it is always possible to recognize adequately complete specimens as belonging to one or the other, they are considered to be separate species. It will be obvious at once that this procedure applies to samples of several fossils, only rarely to single specimens, since if the ranges of variation of successive subspecies overlap, it may be impossible to say to which of several subspecies a single specimen belongs. It may be an extreme variant of a population referred to one subspecies, or an example of a common variant in another.

One of the most important accounts of a single gens with considerable variation was given in a classic paper, by A. W. Rowe, who was one of the first, perhaps actually the first, to appreciate fully the rival claims of stratigraphy and evolutionary research, and the use of populations, not selected individuals, as a basis for naming species. His paper appeared in 1899 and is astonishingly modern in almost every way. In many respects he was far ahead of his time, and is behind the times today in only a very few matters.

He set out to examine carefully variation in the gens *Micraster*, an extinct genus of sea-urchins found abundantly in the chalk at Beachy Head and elsewhere. His two purposes are clearly stated. He says that in studying variation in these very variable forms we must either follow certain European palaeontologists who had based each of a huge number of specific names on one or two specimens regardless of intermediates "on the plea that every minute variation must be ticketed and pigeon-holed, irrespective of the fact that such variations may be valueless as zonal guides," or we must "study the facies of the genus as a whole, carving out broad zoological groups, and allowing the horizon and not the species, to be our criterion. In this way, all valid species and varieties will be retained, and those which are valueless as zonal guides will soon find their level, and sink into the oblivion of an unwieldy synonymy.

"To such a scheme as this the writer unhesitatingly gives his allegiance, for in it we may hope to trace the evolution of the genus as a whole, and each prominent feature of the test [shell] in particular, and in it the neglected, but equally valuable passage-form will receive its due recognition."

Very many of the 'species' created in this genus had been based on badly localized specimens. Rowe realized that the mere fact that many of these, plus those intermediate forms (passage-forms) which were frequently disregarded, could be arranged in a perfectly graded morphological series was not enough to establish the course of evolution unless the series was also in true sequence of time. He collected specimens of the genus from the chalk, working over the rock-face inch by inch for many feet, until he had obtained over 2,000 specimens whose exact relations in the time-series had been most carefully determined. He was able to establish that there had been a gradual and progressive transformation in time of the characters of the shells, to demonstrate clearly the relationship between individual variation at any one time and this progressive variation, and to show that provided one's specimens were accurately localized, one could use this genus for stratigraphical purposes, although it had previously been considered as useless since most of the 'species' recognized could occur anywhere in the time-sequence.

The shell of *Micraster* is a somewhat heart-shaped object, bun-shaped in side view, made of numerous small interlocking or fused plates. Many 'species' had been erected solely on its profile, irrespective of any other character. '*Micraster normanniae*', for example, referred to all very flat shells. Such 'species' could occur at any horizon, although with varying degrees of abundance, and in each horizon could be completely interconnected by series of intermediate specimens. They represented merely individual variations of no value stratigraphically, and he reduced them to infraspecific rank as *formae*. Thus in any population 'forma *normanniae*' could be applied to very flattened shells, 'f. *beonensis*' to high rounded ones, and so on.

Other characters, although showing a high degree of individual variation at any one time, showed a progressive change in the time series. Provided that one considered not individual specimens but fairly large samples from each horizon, these characters could be used to distinguish successive forms. Rowe found that there was considerable variation in those plates of the shell that lie between the rows of holes originally connected with tube-feet—the ambulacral plates. He described these, in order of increasing distinctness of the individual plates, as smooth, sutured, inflated, subdivided and divided, giving very good photographs to illustrate his terms. He recognized in the chalk beds he investigated five successive zones. Smooth ambulacra could be found in specimens from any but the topmost two zones. In the bottom two, almost all specimens had smooth ambulacra, in the next only about 20 per cent. The sutured type, with definite slight boundary-lines between the individual plates composing the ambulacra, appeared as an occasional individual variant in the bottom zone but one and was found in 44 per cent of specimens from the next (the middle) zone, but not at all in any of the higher ones. The inflated and subdivided types comprised 30 per cent of the population in the middle zone, 50 per cent in the one above it, while the extreme divided type appeared first in 3 per cent of the top-but-one zone, rose to 20 per cent in the lower third of the top zone, and was the commonest form above.

Similarly, shells from the two lowest zones were more

elongate-ovoid, with only a slight flattening (not a furrow), at the anterior end, and the mouth comparatively rather posterior. All these characters changed progressively, like those of the ambulacral plates, so that shells from the topmost zone were more rounded, with a deep furrow anteriorly, and with the mouth rather close to the anterior edge of the shell. But all varied independently so that a specimen with, for its zone, extreme ambulacra might have a quite normal shape, for example.

At every horizon were found numerous passage-forms, that is, individuals with characters intermediate between those of the named forms occurring in that horizon. As already mentioned, in the middle zone, 20 per cent of his specimens had 'smooth' ambulacra, 44 per cent had the 'sutured' type, and 30 per cent showed the 'inflated' and 'subdivided' types. But these types are not sharply defined, and there were many specimens with ambulacra that fell by definition within his 'sutured' class that only just showed the suturing and were really intermediate between 'smooth' and 'sutured', while others connected the 'sutured' and 'inflated' types. In fact, variation was continuous at each zone, but the mode of variation shifted progressively, so that while in the zone below 'smooth' ambulacra were by far the commonest, in the middle zone 'sutured' forms were more abundant, and in the uppermost, the 'divided' form.

Specimens of *Micraster* are rare in the lowest zone, common to abundant in the others. By studying the variation in all characters, Rowe was able to show that in each of the four upper zones the commonest forms possessed particular combinations of character which could be used to define them and to recognize the zone. He recognized binomially five main types, distributed as follows:

Lowest zone:	*Micraster cor-bovis.*
Second zone:	*M. cor-bovis*, a very few *M. leskei.*
Third zone:	A few *M. cor-bovis*, many *M. leskei*, a few *M. praecursor* and *M. cor-testudinarium.*
Fourth zone:	Many *M. praecursor* and *M. cor-testudinarium.*
Fifth zone:	Lower third as fourth zone, but the specimens have more advanced characters; rest. *M. cor-anguinum.*

For the three lower zones other fossils were already available as zonal guides. Rowe was able to define the two uppermost ones as the zones of *M. cor-testudinarium* and of *M. cor-anguinum*. But having defined these species and their zones, he was faced with the necessity for admitting that, for example, in the populations in the third zone, referred principally to *M. leskei*, could be found some extreme variants which by their morphological characters must be referred to the earlier *M. cor-bovis* on the one hand, and others referable to the later *M. praecursor* on the other. Since in each zone every possible intermediate between the named forms can be found, it is clear that the *M. praecursor* of the middle zone, say, were no more the sole ancestors of the *M. praecursor* of the next zone than the class of all tall men and women with dark brown eyes and black hair today will be the sole ancestral class of all such men and women ten generations hence.

It is obvious that Rowe was dealing with a single gens, showing considerable morphological variation at any one horizon plus a gradual change in variation in the course of time, as he himself clearly realized. From the biologist's point of view, only one binomial can be applied to any one time-section of the gens, not several; while from the stratigrapher's point of view, Rowe's classification is perhaps the best possible, since it is backed by clear, simple definitions of the named forms and it makes available two zonal guides. The use of a single binomial for the Micrasters of each zone would necessitate using vague definitions covering a great deal of individual variation, and requiring a careful determination of the frequencies of the varieties present, which could be done only if a large number of specimens from each zone or subzone were available. Nevertheless, this must be done, if the species recognized are to be valid biologically. One might call all the specimens from the bottom two zones *M. cor-bovis*, from the mid-zone *M. leskei*, from the fourth and lower third of the fifth *M. cor-testudinarium*, and from the rest *M. cor-anguinum*. That Rowe's classification is still largely arbitrary is obvious from the fact that one could obtain an equally valid, or rather equally arbitrary, classification by naming those variants which Rowe called passage-forms. In the middle zone, for example, *M. cor-bovis*, *M. leskei*, *M. praecursor* and *M. cor-testudinarium* all coexist. One might equally well recognize as species *M. cor-bovis* (say), the passage-forms between

M. leskei and *M. praecursor*, and those between *M. praecursor* and *M. cor-testudinarium*. From the stratigraphical point of view that classification which most readily lends itself to use in the field and to the production of zonal guides is to be preferred.

In *Micraster* the gens can be recognized with some certainty. In many other fossil groups the situation is far less clear because many gentes were in existence at the same time, and often showed the most remarkable parallel variation in certain features. Different species belonging to different gentes but living at the same time have many features of the ornamentation of the shell in common. The phenomenon is particularly well known in the Ammonites. Arkell and Moy-Thomas give as an example the four successive Ammonite genera *Cadoceras*, *Quenstedtoceras*, *Cardioceras* and *Amoeboceras*, each with obvious characteristics, easily distinguished, and of great value to stratigraphers. Each genus contains forms which in some characters are more like members of the other genera than they are like the rest of their own genus. On the basis of such characters, each genus has been split up into subgenera. "As knowledge advances, it is becoming increasingly probable that each subgenus of *Cardioceras* evolved from a different subgenus of *Quenstedtoceras*, either known or still undiscovered. The earliest *Cardioceras* in English deposits, the genotype [type of the (sub)genus] of *Scarburgiceras*, is indistinguishable until the later stages of its development from the last representatives of the latest subgenus of *Quenstedtoceras*, namely, *Pavloviceras*. There is greater 'true' affinity between *Pavloviceras* and *Scarburgiceras* than between either and its companion subgenera; yet *Pavloviceras* is obviously a *Quenstedtoceras*, and *Scarburgiceras* when full grown can only be classed as a *Cardioceras*."

In such an example as this, the resemblances between gentes or groups of gentes at any one time are more striking than the resemblances between the successive portions of a single gens. In consequence the genera most useful stratigraphically are horizontal, not vertical, containing contemporaneous parts of many gentes, not whole gentes as far as known. Such genera are of course polyphyletic; the characters of each have been acquired independently in many separate although closely related lines. Natural groups should always be monophyletic. But the introduction of vertical genera containing whole gentes is often very difficult, either because each gens changes so much that no diagnosis of it can be offered, or more frequently, because the evidence is so incomplete that one cannot be sure of the gentes. The affinities of other subgenera in *Cardioceras* and *Quenstedtoceras* are uncertain. *Vertebriceras* is a subgenus of *Cardioceras* that could have arisen equally well from either of two subgenera of *Quenstedtoceras*, and it is impossible to determine which.

In such situations as these, a compromise is inevitable. It seems to be generally agreed that a genus may well have arisen more or less simultaneously by transformation of several gentes, but its polyphylety can be tolerated provided those gentes are themselves grouped in a single genus. Natural groups must be defined such that no one group can be thought to have arisen from more than one group of the same rank as itself, or of higher rank. (It must be admitted that those examples of plant species which appear to have hybridized and fused completely must form an exception to this rule.) And where the gentes cannot be made out with reasonable certainty, one can only use a morphological classification, which may well be a horizontal one. The employment of strict evolutionary criteria and the delimitation of gentes is usually impossible because of lack of evidence.

INTERRELATIONSHIPS OF THE VARIOUS SORTS OF SPECIES

We can now summarize the interrelationships of the various sorts of species.

(i) The *taxonomic species*, or more simply, 'the species', without qualification. This category includes all other categories of species, and therefore all named species. It can be defined as the category of all concepts based upon one or more specimens and referred to by a binomial which satisfies the requirements of the International Rules of Nomenclature. One must speak of concepts, since even if a species is known only from a single specimen, the species-description based on it is assumed to apply to many other specimens, observable in theory. The specimen or specimens available are treated as a sample of a much greater number of individuals, and the species-description based on this sample is an inference of the distinguishing characters of the whole population. All taxonomic species can be placed in at least one of four categories, of very different status.

(ii) The *morphological species*, or morphospecies. This category contains all those species which have been established solely on morphological evidence.

(a) Species based on specimens which are extreme variants of a single morphological series, such as *Micraster normanniae* (p. 116). These are not allowed in neontology, and in palaeontology are of use only to stratigraphers. It is probably desirable that this category should be abolished.

(b) Species based on specimens which appear to represent morphological series not known to be continuous with any contemporary series. Such species may in fact be, or have been, sexual or asexual, and sympatric or allopatric with their nearest known relatives. This category contains the great majority of known species.

(iii) The *palaeospecies*. This category includes all species known to intergrade so as to form chronological series, for example, the series of species of *Micraster* given above (p. 117). They are arbitrary sections of gentes, consequently their morphological limits in the time series are also arbitrary. The only real distinction between these and modern species or neospecies is that the latter have a natural, not an arbitrary morphological limit in the present (namely, their morphological characters now), but not, of course, in the past, since they are the temporary terminators of gentes. In practice there is an important distinction, because it is often possible to determine the mode of reproduction used by living forms, but this can only be guessed at in extinct ones. Since no living vertebrate is parthenogenetic, and the mammals are merely one class of vertebrates, it is reasonable to assume that a particular extinct mammal was not parthenogenetic, but this is only a reasonable assumption. In groups such as the Crustacea, where parthenogenesis is frequent, such an assumption might be very dangerous. It is probable that the palaeospecies of very long gentes which show a great deal of evolutionary change reproduced sexually, for the reasons given on p. 104.

(iv) The *biospecies*. This category contains all those species that can be shown to conform to the biological definition of the species (pp. 86, 93). It therefore contains only species with true sexual reproduction, genetically isolated from all other species with which they coexist, for example, the biospecies listed on pp. 73, 75 and 93. Usually biospecies have distinctive morphological characters (p. 75), but these may be subject to extreme geographical variation (so that for the identification of specimens, their locality must be known) or may be exceedingly difficult to detect, or even apparently absent (extreme siblings, p. 80). The criterion of morphological distinctiveness may be useless on occasion, and only the genetical criterion can be regarded as certain. But if there is an obvious morphological gap between a polytypic morphospecies and all its relatives although it is sympatric with some or all of them, then the gap can be taken to indicate a genetical separation also. Consequently many species which are strictly speaking morphospecies can be brought into the category of biospecies provided only that they are believed to reproduce sexually.

The limits of the biospecies are uncertain spatially when allopatric populations are considered, since the genetical criterion cannot be applied, and only comparative studies of morphology, physiology, genetics and behaviour are available. They are genetically and spatially uncertain when secondary hybridization occurs regularly. And they are uncertain when one passes back in time, because biospecies, like all neospecies, are the present terminators of gentes. In practice, the immediately preceding palaeospecies are known in but few examples.

(v) The *agamospecies*. This category contains all those species that reproduce entirely asexually, and those that use the appurtenances of sexual reproduction but without a fusion of gametes. *Eiseniella tetraedra* is an example (p. 99). They can be brought readily into the natural classification, but their exact rank cannot be determined, since they can be classified solely by general similarity or dissimilarity, like morphospecies.

It is obvious that each of the four categories, the morphospecies, palaeospecies, biospecies and agamospecies, relies on a different criterion, and that they are far from mutually exclusive. The morphospecies is an expression of our ignorance. One must have some evidence before a species can be described at all, and the evidence most easy to obtain (in palaeontology often the only evidence that can be obtained) is morphological. Forms recognized as morphological species may in fact be (or have been) biospecies, agamospecies, subspecies of polytypic species, or complexes of sibling species which themselves might include some agamospecies and some sexual ones. If plenty of specimens are available, one may often very reasonably infer that one's morphospecies are at any time morphologically distinct groups almost certainly representing biospecies which are stages in different gentes. In some particularly well-known fossil groups, the status of the taxonomic species may be as well known as in many of the less well worked groups of living organisms.

The palaeospecies is an expression of the attempted imposition of a hierarchy developed for classifying discrete groups, on to continuous evolutionary series. Because of the imperfections of the fossil record many fossils do fall into morphologically discrete groups and can readily be incorporated into the hierarchy. Nevertheless, the whole concept of the species as a morphologically (and by implication genetically) discrete group is based upon observation of present-day animals, and holds only for short periods of time (p. 107) which on the evolutionary scale are mere instants. The palaeospecies is an uncomfortable compromise between theoretical recognition of the continuity of each gens, and the necessity for incorporating fragmentary fossil remains into the natural classification.

The agamospecies represents an advance on the morphospecies, since the mode of reproduction is known. Unfortunately, it is a mode which allows no possibility of framing a definition of the species which is any less arbitrary than the morphospecies. Some agamospecies (like some morphospecies which are believed to be based on sexually reproducing animals) are morphologically very distinct, others are not. Natural groups of individual specimens, and groups of groups, can readily be formed, but indisputable ranks cannot be assigned to them. Indeed, there is always the hope that additional evidence may enable one to show that a particular morphospecies is a biospecies, but this can never happen with an agamospecies known to be totally asexual. Specific rank in agamospecies as in morphospecies has only positional value (p. 102).

The biospecies, on the other hand, has a very different status. Biospecies as such, like animals or individuals as such, are definable, and consequently they are of enormous systematic importance. They are also of the greatest evolutionary interest (Ch. viii). A biospecies can be recognized only after considerable investigation, whereas a morphospecies can be based on a single piece of bone or shell or even a natural cast of such a piece. The evolutionary status of a biospecies can be determined, while a morphospecies, as indicated above can be any one of many very different entities, or even a mixture of them. Biospecies in general retain the power of comparatively rapid response to changing conditions, and are almost certainly the constituents of the greater part of the evolutionary tree, while agamospecies are probably almost always sideshoots perhaps persisting for a long time, but not ancestral to large groups.

In consequence there has been a noticeable tendency in recent years, especially among geneticists who must of necessity confine themselves to living forms, to speak of the biospecies as 'the species' and to deny specific rank to all other sorts of species. The quotation from Dobzhansky on p. 105 shows distinct traces of this attitude, which tends to neglect the functions of the systematist, the limitations of the biospecies, and perhaps considerations of what is practical. The biospecies is a definable concept only if time and allopatric populations are ignored and asexual forms are excluded from consideration. But no palaeontologist can ignore time. No student of speciation can avoid the problem of the status of allopatric forms. No student of evolution can ignore agamospecies. And no systematist can ignore any forms within the groups he works on merely because they are asexual.

The problems involved in any large-scale reformation of the present system are best shown by means of a fictitious example, demonstrated in figure V. This diagram is not a simple evolutionary tree. Time advances to the present day up the page, but the other two axes are latitude and longitude, and the piece of 'tree' shown in outline represents the geographical changes in time of a small archipelago. This consisted at the beginning of the period considered of one large and one small island. The small island has persisted more or less unchanged to the present day. The large island, the shoreline of which has fluctuated considerably, was broken for a time into two halves which quickly rejoined; but soon after the junction, one mountain block became a separate island by tilting and is still separate. At the present day five distinct forms, all referred to the genus A, inhabit these islands. These are the parthenogenetic form *Aa*, commonest in the drier western region of island X, and two intergrading forms,

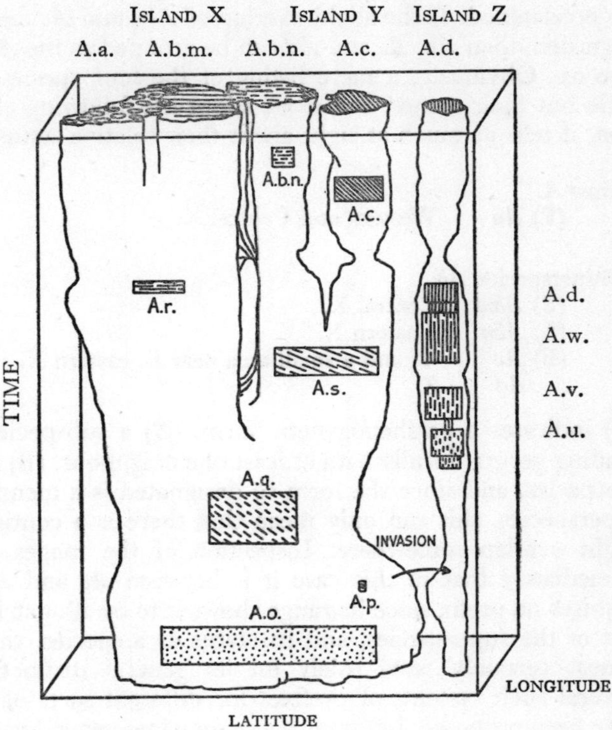

FIG. V. Inter-relationships of different sorts of species

obviously subspecies of a single species, *Abm* and *Abn*, in the central and eastern regions of X. On Y occurs *Ac* which has invaded X apparently very recently, occupying only land disturbed by human activities near the only village on X. Finally, on Z is a rather distinctive form *Ad*. The different forms of shading in the plane of the present day (which is a map of the archipelago at present) indicate the different morphological characteristics of these animals and the extent of each sort of shading gives their geographical distributions. At the present day, then, there are four (taxonomic) species, *Aa*, *Ab*, *Ac*, *Ad*. *Aa* is an agamospecies. *Ab* is polytypic, and is a biospecies at least at the eastern extremity of its range, with respect to *Ac*, since *Ac* and *Abn* overlap without hybridization. *Ad* is allopatric to all the others, and its exact status with respect to them cannot be determined. If the slight overlap of *Abn* and *Ac* can be disregarded, then *Ab*, *Ac* and *Ad* can be grouped as the *Ab* superspecies. Obviously, a mere listing of the four species tells us little but their names. But if we adopt the following classification, it tells us much at sight about their relative status.

Genus A.
(P) *Aa* Western and Central X.

Superspecies *Ab*
(S) *Abm* Central X.
(S) *Abn* Eastern X.
(B) *Ac* Y, and a small area near L, eastern X.
Ad Z.

(*P*) indicates a parthenogenetic form, (*S*) a subspecies intergrading geographically with at least one neighbour. (B) means a biospecies, and since the form so designated is a member of a superspecies, this can only mean that there is a contiguity or slight overlap somewhere. Inspection of the ranges tells us immediately that in this case it is between *Ac* and *Abn*. *Ad* requires no prefix since its range shows it to be allopatric to the rest of the superspecies, and its status as a species, therefore, almost certainly rests solely on its general distinctiveness. Several such systems of prefixes for different sorts of species have been proposed, but none have come into general use as yet.

Now suppose that from deposits on the islands, a number of samples of fossils are known. The shaded rectangles in the diagram represent these. The position and height of each relative to the time scale give the time relations of each sample. The width is proportional to the number of specimens in each, and is an indication of how adequately each sample can be taken to represent the population from which it originally came. The shading of each rectangle indicates the morphological characteristics of each population. Taxonomic species have been based on all these samples except those morphologically identical with living forms. (As usual, many were erected on single specimens and are now sunk as synonyms, and only the valid names are considered here.)

All these fossil species are morphospecies, as indeed is *Ad*. Four intergrade chronologically and morphologically and must be regarded as palaeospecies of a gens, terminating at present in *Ad.* The other are based on isolated samples. *Ao* seems to represent a stock from which all subsequent forms may have descended. The status of *Ap* is far from clear. Morphologically it is intermediate between *Ao* and *Aa*, but the sample is only of a single specimen. *As* is presumably ancestral to *Ac*, and since there is only one sexual species on Island X presumably it and *Ar* were only subspecies, and interbred on meeting.

The general features of a natural classification of these forms are obvious from the diagram. *Ao*, *Aq* and the problematical *Ap* are more primitive; after them two gentes are distinguishable, *Au*, *Av*, *Aw* leading to *Ad*, and *As* and *Ar*, or more correctly *Ar* with two subspecies, leading to *Ab* and *Ac*. The origin of *Aa* is unknown. The taxonomic hierarchy cannot possibly express their interrelationships adequately; this can only be done by a diagram like figure V, in which both continuity and discontinuity are equally easily shown. Any notation that requires separate symbols or reference names, such as the binomial system, is by that very fact incapable of an adequate expression of chronological or geographical continuity accompanied by morphological change. All the forms can be fitted into a morphological hierarchy, the ranks of which give an indication of the homogeneity of the groups within them (p. 27). In addition *Ab* and *Ac* are recognizable as a biospecies with respect to each other. This means only that their rank in the hierarchy is determinable, and that if they have reached specific rank, then *Ad* should probably be called a species since it is as different from either of them as they are from each other. What rank is given to *Aa* depends on its morphological characters and degree of variation. If it shows little variation and is morphologically as distinct from *Ab* as from *Ac*, it will be called a species by analogy; otherwise it might be only a parthenogenetic race of *Ab*, or if it shows much variation, a subgenus with several species. The recognition of *Ab* and *Ac* as biospecies need have no bearing whatever on the ranking of *Ar*, *As* and *Aw* for example. *Aw* may or may not have been a biospecies with respect to *As*.

It is obvious that all these various forms can be brought together into a single natural hierarchy only if the minimum requirements for classification are satisfied by very badly known fossil forms as well as very well-known living ones. There is no need to wait until we are in a position to make a perfect classification before having any classification at all, because we never shall be able to make a perfect classification. The consistent use of a hierarchy of rank-names enables us to give an indication of the degree of homogeneity to be expected in a group of any particular rank. To call a particular form a species does usually mean that, from the specimens available, it is the smallest possible natural group above those relating to individual variation or intergrading geographical variation. Provided one remembers that such groups may belong to four different sorts of species, of very different evolutionary significance, no harm is done. To restrict the term species to biospecies destroys the uniformity of the natural hierarchy quite needlessly without offering any alternative system. But of course it is most important to realize that no systematist can be satisfied with purely morphological criteria in any group where it is practicable to obtain more than purely morphological data, since morphology *as such* must be treated as only a clue to affinity by descent in all kinds of species, and, in addition, to the genetic limits of the species where biospecies are concerned.

To sum up, four sorts of species can be recognized. *Biospecies* consist of genetically intra-connected populations representing the latest form in a gens. Their limits are indistinct in time, in space where allopatric forms are concerned, and genetically when hybridization is of importance. Each biospecies is definable not by morphological characters or presumed evolutionary affinities but by the (actual or potential) genetic contribution of all its individual members to the production of the next generation. It is the only sort of species that can be defined as such. *Palaeospecies* are sections of a single gens, arbitrarily separated from each other and given specific status on morphological characters. Considered at any one instant in time, a palaeospecies is a biospecies. *Agamospecies* are forms with no sexual reproduction. They are ranked as species on purely morphological grounds. *Morphospecies* are forms given specific rank on purely morphological grounds, usually too incompletely known to be assigned to any of the preceding sorts of species. (Morphospecies for which there is good indirect evidence that they reproduce sexually and coexist with their nearest relatives without intergradation are by inference biospecies.)

At every point where other criteria fail, one must fall back upon comparative studies, in theory embracing all aspects of the organisms concerned, but in practice almost entirely morphological. The accepted taxonomic system may be described as a natural system based upon comparative studies of individual specimens, corrected where possible by reference to the genetical-evolutionary criteria, and employing discrete groups because the discreteness of most natural groups at the present day, and the imperfections of the fossil record have as yet almost entirely obviated the necessity for recognition of the essential evolutionary continuity of all living things. Its special virtues are that the addition of new forms does not mean a complete reorganization of the whole classification and that the qualifications for entry are so low that imperfectly known forms can be admitted (with a corresponding degree of uncertainty about their precise position in it.) In consequence of this latter property, and since morphological similarity, with certain important safeguards (p. 75) can be taken to mean evolutionary affinity, such forms as agamospecies can be placed in the natural classification next to their closest relatives, sexual or asexual, where from the evolutionary point of view they undoubtedly belong.

CHAPTER VIII

GEOGRAPHICAL SPECIATION

ONE has only to imagine the consequences of unrestrained hybridization between all living animals to appreciate the extreme importance of the process of speciation. If all specific barriers were suddenly and miraculously removed, the result would be an appalling welter of hybrids with every possible combination of characters. No single individual would be properly adapted to any one mode of life, and many of the characters of each might be adapted for well-nigh incompatible purposes. While most of the possible combinations could not survive for long and would die out, the few practicable ones, by continual cross-breeding, would settle down to a single 'general purpose' species, able to maintain itself but not adapted to do any one thing with special efficiency.

The various sorts of barriers discussed below (pp. 160, 162) which prevent wholesale hybridization enable different species to evolve in genetical independence of each other, each specializing for a different mode of life as a swimmer, burrower, jumper or flier, a sucker, chewer or gnawer, a deep-sea form, an internal parasite, an inhabitant of hot springs, or some other of the half-million or so trades and professions followed by different species of animals. Without such barriers the extraordinary specializations that we see in different animals (and plants) would be impossible and the whole evolutionary tree would consist of a single unbranched twig.

It is only within the last twenty years that some of the processes involved in producing new species have been clearly described. Darwin himself, inclined to place more emphasis on the species as merely one rank in a continuous series than on its properties as such, did not appreciate some of the difficulties involved, although he certainly realized and stated explicitly the importance of geographical variation. In his autobiography he writes: "During the voyage of the *Beagle* I had been deeply impressed by discovering in the Pampean formation [of South

America] great fossil animals covered with armour like that on the existing armadillos; secondly, by the manner in which closely allied animals replace one another in proceeding southwards over the Continent; and thirdly, by the South American character of most of the productions of the Galapagos archipelago, and more especially by the manner in which they differ slightly on each island of the group; none of the islands appearing to be very ancient in the geological sense.

"It was evident that such facts as these, as well as many others, could only be explained on the supposition that species gradually become modified; and the subject haunted me. But it was equally evident that neither the action of the surrounding conditions, nor the will of the organisms (especially in the case of plants) could account for the innumerable cases in which organisms of every kind are beautifully adapted to their habits of life—for instance, a woodpecker or a tree-frog to climb trees, or a seed for dispersal by hooks or plumes. I had always been much struck by such adaptations, and until these could be explained it seemed to me almost useless to endeavour to prove by indirect evidence that species have been modified."

This passage is of the utmost importance in explaining Darwin's thoughts on geographical variation. The *fact* of such variation—the fact that in the Galapagos Islands each island had its own endemic form of tortoise and of many birds, each distinct but obviously related to forms on other islands, was startling. Why should each island have its own special form? The remarkable variation in the beaks of the different forms of the Galapagos finches, all obviously closely related, and all confined to the Galapagos archipelago, made an impression on him that deepened with time. In the second edition of his *Journal of Researches during the Voyage of H.M.S. Beagle*, better known under a later title *A Naturalist's Voyage round the World in H.M.S. Beagle*, he says "Seeing this gradation and diversity of structure in one small, intimately related group of birds, one might really fancy that from an original paucity of birds in this archipelago, one species had been taken and modified for different ends." Lack has remarked, "This last phrase is the most significant in the whole book, and is Darwin's first public pronouncement on a subject the elaboration and generalization of which was to occupy the next fifteen years of his life."

The fact of geographical variation and representation, then, was principally responsible for his recognition that species were not immutable. But also he had been even more deeply impressed by the remarkable specializations of animals which adapt them for particular modes of life, and for the origin of which some explanation must be found.

"I soon perceived," he says, "that selection was the keystone of man's success in making useful races of animals and plants. But how selection could be applied to organisms living in a state of nature remained for some time a mystery to me." An answer came to him after reading Malthus's essay; there will be competition, and the better adapted forms will be favoured.

Darwin's book *On the Origin of Species* first appeared in 1859. In 1868, 1870 and 1875 Professor Moritz Wagner pointed out in a series of essays based on very extensive research in many parts of the world that new species can arise only through geographical isolation. He showed how universal geographical variation is, and emphasized that if a population became separated from the rest of the species it could diverge in isolation and eventually become a different species, but otherwise continual interbreeding with the rest of the species would prevent it from diverging far enough. Wagner was not the first to entertain such ideas, but he was the most considerable worker on the subject, and the first to point out its importance to Darwin. Darwin wrote to Wagner, in 1876, in his usual friendly style, to thank him for copies of his essays, saying, "I wish, however, that I could believe in this doctrine [of geographical speciation], as it removes many difficulties. But my strongest objection to your theory is that it does not explain the manifold adaptations in structure in every organic being—for instance in a Picus [woodpecker] for climbing trees and catching insects—or in a Strix [owl] for catching animals at night, and so on *ad infinitum*. No theory is in the least satisfactory to me unless it clearly explains such adaptations. . . . I do not believe that one species will give birth to two or more new species, as long as they are mingled together within the same district. Nevertheless, I cannot doubt that many new species have been simultaneously developed within the same large continental area; and in my *Origin of Species* I endeavoured to explain how two new species might be developed, although they met and intermingled on the *borders* of their range. It would have been a strange fact if I had overlooked the importance of isolation, seeing that it was such cases as that of the Galapagos Archipelago, which chiefly led me to study the origin of species. In my opinion the greatest error which I have committed, has been not allowing sufficient weight to the direct action of the environment, i.e. food, climate, etc, independently of natural selection. Modifications thus caused, which are neither of advantage nor disadvantage to the modified organism, would be especially favoured, as I can now see chiefly through your observations, by isolation in a small area, where only a few individuals lived under nearly uniform conditions."

Later, in a letter to Professor Semper, Darwin wrote: "As our knowledge advances, very slight differences, considered by systematists as of no importance in structure, are continually found to be functionally important. . . . Therefore it seems to me rather rash to consider the slight differences between representative species, for instance those inhabiting the different islands of the same archipelago, as of no functional importance, and as not in any way due to natural selection. With respect to all adapted structures, and these are innumerable, I cannot see how M. Wagner's view throws any light, nor indeed do I see at all more clearly than I did before, from the numerous cases which he has brought forward, how and why it is that a long isolated form should almost always become slightly modified."

These passages clearly show what Darwin's difficulties were. The geographical variation shown by Wagner affected characters which *seemed* to have no adaptive significance. Several examples have already been discussed in this book. It seems very difficult to see what possible selective advantage it is to the great tits (p. 55) to be white-bellied and grey-backed in India, white-bellied and green-backed in Japan, and yellow-bellied and green-backed in Europe. The variation described in the rosella parrots (p. 62) seems equally purposeless. Variation in the sculpturing of beetles' wing cases, or of snail shells, or in the size and position of spots on butterflies' wings, and in many other such characters has no obvious significance. It cannot, therefore, lead to the development of adaptations, such as are produced by competition between individuals and species, but might be produced by the direct action of the environment. In the last letter quoted, it is very curious that Darwin should explain so admirably why it is dangerous to conclude that slight specific differences are non-adaptive, and then fail to apply his own precepts to the study of geographical variation. Two other difficulties were the production of new species in large continental areas, and the reasons why isolated populations should diverge at all.

All these difficulties have now been overcome, and it is recognized that geographical speciation is perhaps the principal method of speciation in animals, although certainly not the only one. To summarize the present attitude very briefly, spatial isolation is necessary for sufficient divergence to take place, there are several reasons why isolated populations will diverge genetically from each other, and a better knowledge of geographical variation shows both that much of it is adaptive or is correlated with adaptive features, and that it affects probably all characters of the organism, not only the most obvious adaptations and specializations. Darwin's over-insistence on the importance of obvious adaptation led him to disregard apparently non-adaptive characters and so to attach less than its due importance to Wagner's theory.

EVIDENCE FOR GEOGRAPHICAL SPECIATION

(i) *Relative status of allopatric forms.*

The principal evidence for geographical speciation is, of course, indirect (since the process is too slow to be observed in a few human generations) but it is very extensive. In probably all the best-worked groups of animals geographical variation is found to be the rule, rather than the exception; and every degree of morphological difference between geographically representative and obviously closely related forms can be found. Many isolated populations are, as far as can be seen, identical in all characters. Others differ only slightly. Yet others are sufficiently different to be named as subspecies, while others might be called species—and so on. Some very aberrant representatives have even been separated generically. The same sort of situation occurs over and over again, in birds, mammals, reptiles, amphibia, butterflies, moths, beetles, bees, scorpions, and snails, and will almost certainly be recognized as widely in the less well-worked groups when more information is available.

It is hardly possible to resist the conclusion that the different forms in each series of geographical replacements are local products of their regions that have acquired their distinctive characters in isolation.

Examples of geographical variation in the wren, the great tit, the coconut lorikeet (*Trichoglossus*), two superspecies of rosellas (*Platycercus*), the golden whistler (*Pachycephala pectoralis*), the small fruit-pigeons (*Ptilinopus*), and two nuthatches (*Sitta*) have been given in varying detail above; the list could be extended almost indefinitely. At any one locality there is hardly ever any doubt of the status of coexisting forms. There is no question, for example, that the crimson rosella and the common rosella (*P. eximius*) are good species. The gaps between sympatric forms are well-nigh absolute. But every conceivable intermediate between species in different genera and separate populations of the same subspecies, can be found in series of allopatric forms. The obvious inference from an enormous number of examples is that species arise from geographically isolated populations, not sympatrically, in coexistence with their parents.

Before going on, two comments must be made on this inference. (a) It does not by itself mean that sympatric speciation in some sense cannot occur. In fact one process leading to sympatric speciation is known (p. 178 below). But it does mean that geographical speciation is common and important. (b) Speciation is taken to mean the multiplication of species. But in the production of a gens, discussed above (p. 111), one species is transformed into another, and this into a third, and so on, without any multiplication of the number of species contemporaneous in any period by the production of specific barriers to intercrossing. Many authors use the word speciation to cover both processes—the *multiplication* of species at any one time, and the *transformation* of species in time since they do not differ fundamentally (p. 168). Rensch refers to the first process as *cladogenesis*, the second as *anagenesis*. Other authors call the first speciation, the second phyletic evolution or a similar term. The word speciation is taken here to refer to *cladogenesis*, the multiplication of species.

The evidence for geographical speciation does not rest only on numerous examples of allopatric series. Examples of invasion are in some ways even more instructive, and can be conveniently discussed under three headings, simple overlaps, multiple invasions, and ring species.

(ii) *Simple overlaps.*

Many series of closely related forms are completely allopatric. Sometimes a whole genus or subgenus may consist of a single polytypic species or superspecies. For example, there are, in the latest check-list ten species of white cockatoos (*Kakatoe*) in five subgenera, of which three subgenera contain only a single polytypic species. The others, with two and five species both contain a single superspecies.

A nearly continuous series of examples could be constructed, to illustrate the passage from wholly allopatric groups of species to groups where all the species overlap more or less completely. One intermediate example has been given above. There can be no doubt that the four species of the *Platycercus eximius* superspecies have arisen by isolation in the four main refuges of Australia during the recent arid period. With the slight subsequent alleviation of the climate, two (*P. adscitus* and *eximius*) have spread, met, and overlapped without interbreeding, thus demonstrating their specific distinctiveness. The closely related genus *Barnardius* survived in the south-western and south-eastern refuges. The south-western refuge became somewhat drier than the south-eastern, and several of the forms in it, having been forced to adapt themselves to drier conditions, have spread far to the east, while the south-eastern forms have remained confined to damper regions. In this way *Barnardius zonarius* has spread from the south-west into mid-western and central Australia, and has encountered the eastern representative (*B. barnardi*) north of Adelaide, where the two seem to hybridize, showing that specific isolation is not yet completed.

The mangrove kingfisher, *Halcyon chloris*, and its relatives show two overlaps. *Halcyon chloris* has an enormous range, from Abyssinia to some islands of Samoa, and is replaced from the rest of Samoa to the Marquesas by forms usually separated as a distinct genus (*Todirhamphus*). The sacred kingfisher, *Halcyon sancta*, is a close relative, found in southern Australia, which must have arisen as a geographical representative. *H. chloris*, however, has extended its range down the eastern coast of Australia well into the breeding range of *H. sancta*. In the Caroline Islands, to the north of New Guinea, is another representative of *H. chloris*, namely *H. cinnamomina*. This has spread westwards to the Palau Islands, and on Palau it now co-exists with members of *H. chloris* forming part of an extension of the range of that species from the Molucca region north-eastwards via the Palaus to the Marianas.

In Europe there are many pairs of very closely allied species, and it has been suggested that most of them arose by geographical isolation due to the ice age. At the time of greatest expansion of the ice there was only a comparatively narrow ice-free corridor across central Europe between the Swiss and the northern ice-caps. It seems likely that many animals requiring more temperate climates were gradually driven south-west and south-east, and persisted in the Iberian peninsula and the Balkans. As the ice retreated, they were able to spread north again, and have now met and show various degrees of hybridization, or specific distinction with or without extensive geographical overlap. Among birds, the western and eastern nightingales (*Luscinia megarhyncha* and *L. luscinia* respectively) meet in Central Europe but overlap only slightly. In the zone of overlap they show preference for different types of vegetation. The newts *Triturus cristatus* (western) and *T. marmoratus* (eastern) show a similar pattern of distribution. They overlap slightly in central France, and in this region a few hybrids are occasionally found. On the other hand the two tree-creepers described above (p. 77), *Certhia brachydactyla* (western) and *C. familiaris* (eastern) now overlap widely since one has become a mountain form, the other a lowland one, and they seldom compete. The two common newts *Triturus vulgaris* and *T. helveticus* are similarly distributed, with a wide zone of overlap in Britain and west-central Europe.

Obviously, if two species arise in isolation but invade each other's territory completely, it will be impossible to say where each originated. In New Guinea there is a very rich avifauna, with many groups of very closely related forms. It is known that much of New Guinea is, geologically speaking, rather recent, and that there have been great changes in the distribution of sea and land. It is probable that the island has been broken into several islands and reconstituted, perhaps more than once. Many of the very closely related groups of coexisting species may owe their origin to geographical isolation followed by interpenetration of ranges, rather than to the survival in this large and ecologically very diverse island, of very many forms which died out elsewhere. The closely related fruit-doves *Ptilinopus ornatus* and *Pt. perlatus* may well have arisen in this way.

In some genera this process can be seen in action. The kingfisher *Tanysiptera galatea* has several subspecies in various parts of New Guinea and some of the adjacent islands. One form (*hydrocharis*) is found both on the Aru Islands and in southern New Guinea, where it overlaps, without interbreeding, with one of the mainland forms. It is believed that until comparatively recently, southern New Guinea and the Aru Isles formed a single island isolated from the rest of New Guinea by a trough of the the sea. When southern New Guinea became joined to the rest of the main island it was colonized by *T. galatea minor* from the south-east, which now coexists with *T. hydrocharis*. Expansion of ranges may still be going on, or—more likely—special habitat-preferences are limiting further spread.

Many other such recent expansions are known, some producing successful coexistence, others only hybridization. One subspecies of *Pachycephala pectoralis* has spread from the north of Australia to New Guinea and now overlaps with the local representative, *P. soror* (p. 64). However, a different one which has apparently arisen on (one or more) small islands and is now colonizing only among the small islands, has extended its range from the Bismarck Archipelago into the Solomon Islands. The subspecies of the Solomon Islands are very distinctive since the males have a yellow, not a white, throat. On one very small island the invaders have met a colonizing population of one of the Solomons forms, and the two have hybridized completely. Similar hybrid populations have been described from several of the Fijian islands.

Such overlaps as those described above show clearly what can happen when previously isolated forms meet, and how sympatric species may be produced.

(iii) *Multiple invasions.*

Multiple invasions occur when the same stock colonizes an isolated district (usually an oceanic island or a mountain-peak) several times, producing several species in it. These differ from simple overlaps only in that each separate invasion produces a population cut off from the parent stock, and so able to speciate in isolation. If it has diverged far enough, this population will not interbreed with the next lot of invaders. If, however, there is continual slight invasion, the resulting gene flow will prevent the island population from becoming very distinct; consequently continental islands do not show this phenomenon (unless they possess high mountains which may be colonized by forms from *distant* mountains on the mainland).

Very many examples of multiple invasions are now known. The European chaffinch has invaded the Canary Isles twice, producing first the blue chaffinch *Fringilla teydea*, and then forms which have not yet acquired sufficiently distinctive characters to be given more than subspecific status within the limits of the common chaffinch (*Fringilla coelebs*). The genus *Ptilinopus* has invaded Fiji three times and the Marquesas twice. Norfolk Island has been invaded three times by *Zosterops*. Such multiple invasions are of the greatest evolutionary interest.

The triple invasion of Fiji by *Ptilinopus* has only recently been recognized as such, since the earliest products are so distinctive that they have been placed in a separate genus (*Chrysoena*). The most recent invasion has produced in Fiji, Tonga and Samoa the form usually given specific rank as *Pt. porphyraceus* (with various subspecies on different islands). It is one representative of a huge series of very closely related allopatric forms beginning with *Pt. coronulatus* of New Guinea, *Pt. monacha* in the North Moluccas, and *Pt. regina* in Australia and the Lesser Sunda Isles, and stretching eastwards to the Marquesas (where there has been a double invasion) and northward to the Carolines, Palaus and Marianas. *Pt. porphyraceus* is merely a not very distinctive local member of this species-group, which is often called the *purpuratus* group, since *Pt. purpuratus* is that member of it which bears the oldest valid name. (There are in fact no rules about the names of species-groups, and there is no reason to take the oldest name, which may well be borne by that form which is assigned to the group with least certainty.)

Also in Fiji and Samoa is a remarkable fruit pigeon, *Pt. perousii*, obviously closely related to the *purpuratus* group, but with unique characters. Instead of being principally dark green and dull grey, the male has much white and pale yellowish-green, with a pink bar across the lower breast and a deep red stripe across the back. The female is much duller, without the pink and red, and much more like a normal member of the *purpuratus* group, in which (except for one isolated form) the males and females are alike. The striking characters of *Pt. perousii* indicate that it is a much older form than *Pt. porphyraceus*. On the other hand, the fact that it occurs on many islands of the Fijian, Tongan and Samoan archipelagos with only slight subspecific differences must mean either that it has spread within the area recently, or that there are rather frequent migrations of individual birds or flocks from island to island.

Lastly, on the three most mountainous islands of Fiji only are three remarkable species, which form a single superspecies. Here again there is strong sexual dimorphism. The females are all dull dark-green birds with yellow beneath the tail, and greenish yellow on the head and belly. Their resemblance to females of various species of *Ptilinopus* (not in the *purpuratus* group) is strong enough to show what their affinities are, but not strong enough to show from which species-group of *Ptilinopus* they arose. The males are very different. One is bright flame-colour with a honey-coloured head; its plumage is very soft and fluffy. One is bright greenish yellow above and pure yellow beneath; in this the feathers are extraordinarily long, thick and strap-shaped, rather like the hackles on the neck of a cockerel. The third is dark green with a yellowish green head, and with some feathers hackle-like, others soft and woolly. These three must be regarded as the result of a very old invasion which itself has speciated in geographical isolation on the three principal islands of Fiji.

In such an example as this just given, the forms resulting from the separate invasions are very distinct and would have been given specific rank even if no overlap were known. In other cases, this is not so, as has been pointed out already for *Pachycephala soror* and *P. pectoralis*. A particularly fine example comes from Tasmania. The thornbills are a genus of small rather dull-coloured birds with several species in Australia. One of these, *Acanthiza pusilla*, is widespread on the Australian mainland and has many subspecies, including one, *A. p. diemensi*, in Tasmania. But also in Tasmania is a form coexisting with *A.p.*

diemensi which in many ways is more like it than it is like some of the other subspecies of *A. pusilla* on the Australian mainland. This second Tasmanian form, *Acanthiza ewingi*, must be given specific status, because on all biological criteria it is a good species. Nevertheless it is very like *A.p. diemensi* which itself is very close to the nearest subspecies on the mainland, which, however, intergrades smoothly with very different forms. Here is yet another instance of the untrustworthiness of morphological characters for determining specific limits. It seems that the earlier invasion of Tasmania by *A. pusilla* produced populations which in isolation rapidly changed (genetically but not very much morphologically) and were able to maintain themselves separately, without swamping, when the later invasion occurred. On the mainland, however, populations living under very different conditions and attempting (so to speak) to change to very different forms, were and are continually swamped by gene-flow from other regions where selection for different characters is going on, so that new specific barriers within this species could not be built up.

Such multiple invasions show that products of the same stock, if isolated for a sufficient length of time, can behave as good species when fresh invasions take place from the parent stock (itself probably altered also in the intervening period). Continual invasion prevents such divergence. It seems clear that in both birds and other animals there must be occasional outbursts of emigration, sufficiently frequent to reach by chance very distant islands but not frequent enough to prevent speciation going on once these islands are successfully colonized.

(iv) *Ring-species.*

But the clearest evidence of geographical speciation is afforded by ring-species, in which two forms that overlap and behave as good species are nevertheless connected by a ring of subspecies, so that no satisfactory specific separation can be made. One example, in the great tits, has been described above (p. 55). It is only by taking advantage of the hybrid zone in Persia that one can resolve the purely taxonomic difficulty; *Parus major major* and *P. m. minor* are connected by a chain of intermediates, yet they behave as good species where they overlap. Several such ring-species are now known. It was originally shown by Stegmann that the herring gull (*Larus argentatus*) and the lesser black-backed gull (*L. fuscus*) which in western Europe are undoubtedly good species, are nevertheless interconnected in this way. There are subspecies of *L. argentatus* in North America which serve to connect it with East Siberian forms, these with West Siberian ones, and these with *L. fuscus*. Several side-chains hang down from this circumpolar ring, producing forms in California, Mongolia, the Caspian and Aral Seas, and the Mediterranean, the Azores, Madeira, and Canary Islands. More recently, Stresemann and Timofeeff-Ressovsky have claimed that the situation in Western Europe is far more complex than this. The north-west was reinvaded four times when the ice retreated, once by *L. argentatus* coming east from North America, once by *L. fuscus* coming north from the end of the Mediterranean side-chain, once by the east Siberian forms coming westward, and once by another expansion of the central-Asian-Caspian-Mediterranean series, advancing north-westwards along the great river valleys of western Russia. Whatever the exact situation, it is certain that all these various forms are interconnected by populations geographically and morphologically intermediate, and that although the ends of the chains overlap and behave as good species, there is no reason to break the chains between one pair of adjacent subspecies rather than between another. Consequently, one must either recognize the specific status of the end forms by making an arbitrary break in the chain and deciding that all forms in one direction must be given one specific name, and all the rest another, or one can avoid this by not making a break, but then good species will have to bear the same specific name. Obviously what is required is a notation additional to the formal taxonomic names, to mark such situations.

In ring-species, it is clear that the effect of distance is to reduce gene-flow between the extreme forms to such a point that they can develop sufficient diversity to coexist without interbreeding. This effect cannot be expected in species with a very restricted range, within which any individual has a good chance of meeting any other, nor in widespread species in which there is very considerable migration so that one individual may breed in a locality far distant from its birthplace. To develop a ring requires also the ability to flood rather rapidly into areas newly available for colonization, so that the end forms do not merely split off and become separate species, as in multiple invasions. Where ring species occur, they provide the finest demonstration of the progressive change of whole populations with distance, resulting in the attainment of specific status—that is, of geographical speciation.

There seems no doubt, then, that geographical (or better, spatial) speciation is a process of very widespread occurrence. Three important questions now arise. How do isolated populations diverge from each other? When they meet, how do they maintain their specific distinctiveness? And how can they invade each other's ranges and coexist ecologically?

PRODUCTION OF DIVERGENCE

(i) *Random incidence of mutations.*

The origin of all diversity in living organisms is, in the last analysis, due to mutations. But most of the inherited diversity actually seen in different populations is due not so much to the appearance of one mutation here and another there, as to the continual shuffling and recombination by true sexual reproductions, of different mutations (pp. 103, 179). It is quite possible that in a number of small populations a particular mutation may well occur in one but not in others. In large populations, with many individuals, it is much more likely that in course of time it will appear in all. Mutations, to a certain extent, are random, since it can only be predicted that a particular one will occur once in a million individuals (say), not in which individual it will occur; and the effects produced by a mutation seem to bear no relation whatever to the needs of the individual that experiences it[1]. The random incidence of different mutations in small isolated populations may play some part in producing diversity. This is especially likely when the mutation consists of a considerable structural alteration in a particular chromosome, such as are known in wild populations. If an alteration of this nature can persist and spread in a population it may play a

[1]It is true, however, that in some organisms the rate of mutation of particular genes is known to be in some degree controlled by other genes. And it is also true that only those mutations can be observed which are not so disastrous that they kill the individuals affected before they hatch or are born. Which are fatal is presumably determined by the constitution of the animal, so that mutations actually seen are to some extent selected by the animal.

large part in making the chromosomes of the population structurally incompatible with those of another, so that if the two populations are able to increase their ranges until they meet, they may have the greatest difficulty in producing hybrids. It can be expected, then, that isolated populations may gradually diverge through differential incidence of mutations.

(ii) *Genetical drift.*

A second process leading to divergence under certain conditions, is known as genetical drift. The theory of it has been worked out by the population geneticist Sewall Wright. It may be described very briefly as the cumulative effect of random sampling errors. Suppose that in a large population of animals there are two contrasting inherited characters—albino, for example, in contrast to normal coloration. Then if we know the frequency in this population of the gene for normal colour, and its alternative, or allelomorph, for albino, it is easy to calculate the proportion of albinos to normal individuals that will appear in the next generation (assuming negligible selection and random mating). If the population is very large, the observed proportion should come very close to the calculated one, because although a landslide or some other accident in one area may by chance have killed off a high proportion of albinos that happened to be there, in another area there may be an abnormally high production of albinos, since by chance several may have met there and mated with each other, not with normal individuals. Again, in one part of the range a mutation may exist which produces deleterious effects when in albino animals; but elsewhere the albino allelomorph may well have met with combinations of other genes that favour it. In very large populations the disturbing factors will tend to cancel out.

Similarly one knows that if a penny is tossed a million times, then it will come down heads very nearly half a million times, since the chances of heads or tails are equal. But if it is tossed only ten times, the equality will not normally be observed—indeed occasionally a set of ten tosses will result in all heads or all tails. Similarly, in small populations the effects of disturbing factors will not cancel out, but will produce considerable divergences from the expected ratio of different genes. But there is this important difference between populations and pennies. Pennies don't breed. The result of any one set of ten tosses is independent of that of any other. But if a small population produces by chance an excess of one form, then that excess is available in the population and may tend to maintain the excess again in the next generation.

In consequence, under certain conditions it is possible that in a large number of rather small populations, all originally containing equal proportions (say) of two alternative forms, by the cumulative effect of random sampling errors, one form may be gradually eliminated in some populations and the other in others, so that diversification will proceed without any effect of selection. When Sewall Wright first put forward the theory (1931) he suggested that it might well account for the production of many subspecific or even specific characters that at that time were believed to be non-adaptive.

In a long series of papers, Sewall Wright has studied the interaction of genetic drift, population size, mutation, and natural selection, and has reached very important conclusions. He has shown that, as would be expected, in very small populations of the order of a few tens or hundreds of individuals, inbreeding is intense and much variability is lost because all of the few individuals carrying a mutation may chance to be killed off in one generation. Consequently "In too small a population there is nearly complete fixation, little variation, little effect of selection and thus a static condition, modified occasionally by chance fixation of rare mutations, leading inevitably to degeneration and extinction." (A gene is said to be fixed when it occurs in every individual in a population, to the exclusion of its alternatives.)

In large populations, on the other hand, genetic drift cannot take place and selection is the principal agent producing changes. But "in a large population, subdivided into numerous partially isolated groups, both adaptive and non-adaptive differentiation is to be expected" and such conditions are most favourable for evolution, since the maximum hereditary variety will be maintained and there is the greatest chance of producing gene-combinations of high selective value.

Sewall Wright's equations and the curves derived from them clearly show the relative importance of natural selection and genetic drift. The simplest measure of selective advantage is that if the numbers of offspring surviving to maturity produced by animals of varieties A and B respectively are in the ratio $\frac{1}{1-K}$, then K is the coefficient of selection in favour of A. It can have any value from +1 to 0. It is often expressed as a percentage (by simply multiplying by 100). To take only one example, Wright has shown that in populations of effective size 10,000, two allelomorphs continually replenished in the populations by mutations from one to the other at a mutation rate of 10^{-6}, will be found in frequencies almost entirely determined by selection if the selection coefficients are as low as 1 per cent or even less, and that the effect of genetic drift is unimportant until the coefficients are reduced to 0.1 per cent or below. Nevertheless, given sufficient time and genes near enough to selective neutrality (0.0 per cent), fixation can occur in completely isolated populations of up to 250,000 individuals, provided that the mutation rate is only 10^{-6}, and in even larger ones if it is less.

The amazing power of selection, even of so low an intensity as 0.5 per cent, to dominate almost completely the frequency of the genes concerned was pointed out in 1930 by R. A. Fisher, and is one of the most important discoveries in evolutionary genetics. It is a most complete confirmation of Darwin's remarks quoted above, on the danger of believing apparently trivial characters to be of no functional importance and in no way due to natural selection. Wright's results also show that selection coefficients must have a very low value before drift can be of importance.

Just what evolutionary importance should be ascribed to genetic drift is a matter of considerable controversy. On the one hand, the process is mathematically certain to occur, given the right conditions, and may even, as claimed, produce by a non-selective process particular combinations of genes which happen to be highly advantageous and can spread, once they have originated, by natural selection. On the other hand, selection coefficients actually determined are all (or almost all) much too high to allow drift to occur, even although they refer to apparently very trivial characters. And it may very well be doubted whether under the changing conditions normal in nature any gene can possess, except for a very short time, a selection coefficient near enough to neutrality for drift to become important in determining its distribution. Genes at a high selective advantage in damp climates may be heavily disadvantageous in dry ones; those that spread in populations of low density may be eliminated under conditions of overcrowding. Many selection coefficients so far observed are exceedingly sensitive to slight changes in the environment; abnormally dry or hot or overcrowded conditions, or slight changes in the nature of available food, may have enormous effects on the abundance of particular genes. It seems most unlikely that genetic drift has been of any great importance in evolutionary change.

Several apparently random variations in gene-frequency have been claimed to be due to genetic drift. Many of them have been shown on further investigation merely to have been insufficiently analysed. It is unfortunately true that if a particular worker can find a definite correlation between the occurrence of some character and a property of the environment, then the distribution of that character is not random, and selection may reasonably be suspected. But if he fails to find any correlation, it does not follow necessarily that there is none. Biological systems—living organisms in relation to their environments—are so complex that it is practically impossible to analyse them exhaustively. There always remains the possibility that further analysis will show correlations. It is fatally easy for a student confronted with a puzzling and apparently random distribution of a character to give up, and declare that it must be due to drift. It is equally easy for a student gazing at a tray of museum specimens to say that he can see no reason for the differences between them, therefore they must be random. Since museum specimens are so often merely parts of animals selected only for their permanence, not their functional importance, and even at the very best are always totally divorced from their natural environments and from all possibility of functioning, it is not surprising that, being separated from everything which might hint at correlations, they appear to vary at random. Darwin's warning is of the utmost importance.

It is, of course, true that very many characters appear to have no importance. A whole book could be written on the nature of characters. Perhaps a tentative definition of a character might be "any part, property, or attribute of an organism, that can be considered independently of the rest of the organism". But this requires further qualification. If one animal is darker than another, the other is lighter than the first; but the relative darkness of one and the relative lightness of the other are one character, not two. This example is only too obvious, but there are other, similar ones, which are much less simple. If one animal is much larger than another, but otherwise very like it, then each part, and each organ are probably larger than the corresponding ones in the smaller form. Are the innumerable size differences all to be reckoned as separate characters, or must they all be counted as one—a general difference in size? If it were known that all the organs could vary independently in size, there might be many separate characters. But if, for example, the large size of one individual were due solely to feeding it with excess of thyroid, then all these 'characters' would be simply aspects of one single change; the large size of the pancreas, for example, would be as necessarily correlated in this individual with that of the brain as is the convexity of a curve viewed from one direction with its concavity viewed from the opposite one.

Consequently, one must never forget that the characters used in taxonomy (or anywhere else in biology) are abstractions from organisms in their environments, which can be made in many different ways. If it should be found that characters usually considered as separate and independent are in fact connected, then the artificial nature of their separation must be taken into account. A very careful investigator has studied the hard parts of two races of the ground-beetle *Carabus cancellatus*, and distinguished 166 separate characters. (Even here, of course, he could not exhaust all the morphological characters even of the hard parts. As Mayr has truly remarked, "the number of characters is limited only by the patience of the investigator.") It is unlikely that any two of these 166 characters should be as obviously interdependent as the two aspects of a curve. Nevertheless, he found that there was correlated variation in many of them, and that this variation was often clearly geographical. One wonders whether this correlation was not in fact due to climatic selection acting perhaps on only a few genes which by affecting the hardening process produced a vast number of 'separate' effects. Such multiple effects of single genes are now well known, and such is the complexity of the developing animal that a single change in the physiology of early development may produce a series of widely differing effects, culminating in the appearance of a most heterogeneous collection of 'characters'. The 'character' produced by a particular allelomorph is in fact merely that aspect of its action which happens to be most conspicuous to us. Work on rats has shown, for example, that the genes affecting particular patterns of coat colour also affect the body weight, the dimensions of the skull, the size of various organs, tameness and docility. Other characters that may also be affected range from the length of the tail to the mean duration of pregnancy. Similar results have been found in very different groups of animals. Consequently, it may well be true that some characters, such as the exact nature of the ornamentations (pits, grooves, or wrinkles) on a beetle's back may not in themselves have any direct value to the beetle (although one should be very cautious in making such affirmations). But if they are necessary consequences of some process which is selected for, then their distribution will also be determined by selection.

(iii) *Natural Selection.*

Evidence of the role played by natural selection in producing geographical variation and divergence is very considerable in importance but not in bulk, principally because the detailed study of geographical variation is really very recent. There are four main lines of evidence, stable clines, the ecological rules, the study of overlaps, and (best, but at the moment rare) direct investigations of selection.

(a) *Stable clines.*

It is very usual indeed to find that a particular character varies geographically in such a way that its frequency or degree of expression decreases smoothly, perhaps even from one end of the range of the species to the other. Such smooth character gradients have been called by Huxley *clines.* They may have any direction, and any extent within the range of the species, and may vary considerably in steepness, or even reverse. For example, in the populations of the coconut lory *Trichoglossus haematodus* in New Guinea, taken from west to east, size increases slightly at first, then decreases very considerably. The blue of the head becomes brighter from north to south, while the yellow collar becomes greener and more indistinct from west to east. Such clines, if stable, are direct evidence for natural selection, without which, in the face of continual interbreeding, they could not be maintained. Yet they often affect extremely trivial characters.

For example, in *T. haematodus* in New Guinea, the breast is red, with a dark bar on the tip of each feather. In north-western New Guinea, these bars are broad and black with a purplish gloss. In south-eastern New Guinea they are narrow and dark green, without any perceptible gloss. If they are used at all as such, by the species, they probably form part of a threat-display or some similar activity. Even with Darwin's warning in mind, it is difficult to believe that their clinal variation as such is of any importance to the bird. Apparently something else, with which they are correlated, is being selected. Examples such as this could be multiplied almost indefinitely.

At the present moment, very few clines have been studied in detail, and the stability of many has hardly been tested. Moreover, it has sometimes been claimed that some clines are in fact independent of climatic gradients (especially those in regions where the climates are very badly known). They seem rather to represent the gradual decrease in frequency of a particular character from the centre of its distribution. But if so, what is causing the decrease? Either the character is still spreading, or there is in fact selection against it in the regions of lower frequency. Mayr quotes the case of the guillemot (*Uria aalge*), investigated by Southern. In populations of this seabird from the north Atlantic occurs a 'bridled' variety *ringvia*, with a white line extending round the head from white rings round the eyes, whereas in the more usual form the head is entirely dark brown. The bridled form is quite common in the breeding birds of the Shetlands, and decreases steadily in frequency to the south. Mayr remarks that "it is difficult to see why the gradual decrease from the north to the south in the number of the bridled individuals in populations of the Atlantic murre (*Uria aalge*) should have an adaptational significance . . ." in relation to environmental gradients. But, as already pointed out, difficulty in understanding a situation is no proof that it does not conform to a particular theory; and on the other hand, there is no need to suppose that the bridle is adaptive as such, only that it is (no doubt incidentally) selected for in the north but not favoured in warmer regions. Selection and adaptation are not exactly the same thing, although some recent authors have unfortunately used them as synonyms.

One may conclude therefore, that the existence of many apparently stable clines in natural populations is good evidence for the widespread action of natural selection.

(b) *The ecological rules.*

It has been known for some time that many different species show the same sort of variation under the same ecological conditions. Such regularities of variation are explicable only by the action of natural selection (always provided of course that variation due to direct action of the environment on each individual animal is excluded. Reduced size in several species of snail in a rather barren region, for example, may be due to no more than a general scarcity of food).

Many ecological rules, summing up these trends of variation, have been proposed. Probably the best-known are Bergmann's, Allen's and Gloger's rules, which apply to warm-blooded animals. They state respectively that within a warm-blooded species body size is larger in cooler regions and smaller in hotter ones, that the extremities of the body (tails, ears, etc) are relatively shorter in cooler regions, and that black pigment (eumelanin) is more abundant in wetter regions, brown pigment (phaeomelanin) in more arid ones. Other rules relate to the number of eggs in a clutch or young in a litter, the development of sculpturing on snail shells, and so on. Some exceptions to all the rules are known. Bergmann's and Allen's rules almost certainly have a simple adaptive significance. By decreasing the surface area in relation to the volume of the body in colder climates they reduce heat-loss and make its control easier. No simple explanation has yet been suggested for Gloger's rule; indeed, Mayr believes that the variation in pigmentation as such may have no significance but merely be correlated with some other variation. This seems rather unlikely, since Gloger's rule holds for the most diverse birds and mammals. It would be strange if in all of these very different forms adaptation to drier climates produced exactly the same by-product of pigmentation. Almost no studies on the effects of pigments in absorbing ambient heat or ultraviolet light from the atmosphere have ever appeared. Even a casual examination of different groups of closely related birds suggests that if there is variation in the habitat occupied and in the quantity of bright red pigment displayed, then red is most abundant in birds of forests, reduced in favour of yellow in birds of open country, and lost altogether in birds of the desert. It is particularly interesting that it tends to disappear first from the upper parts in this series.

Undoubtedly, many more ecological rules remain to be discovered. Wallace's remarks can hardly be improved on. "Many of the most curious relations between animal forms and their habitats, are entirely unnoticed, owing to the production of the same locality *never* being associated in our museums and collections. A few such relations have been brought to light by modern scientific travellers, but many more remain to be discovered; and there is probably no fresher and more productive field still unexplored in Natural History. When the birds, the more conspicuous families of insects, and the land-shells of islands, are kept together so as to be readily compared with similar associations from the adjacent continents or other islands, it is believed that in almost every case there will be found to be peculiarities of form or colour running through widely different groups, and strictly indicative of local or geographical influences."

But it is obvious that body-size is influenced by very many factors besides temperature. An ecological rule relating to climate, say, can only be established if a sufficient number of examples can be observed in which subspecies of the same species (i.e. genetically similar forms) while inhabiting much the same niche, in company with much the same species of other animals, do show variation which can be correlated readily with climate. Other differences in the ecological situation, such as the presence in part of the range of an additional predator or competitor, may well influence body-size to such an extent that Bergmann's rule is overridden. But when the broad ecology of many forms is known, such disturbances can be allowed for, and the general correlation of body-size with climate can be recognized. In fact the disturbances may themselves give an indication of yet other rules. Detailed analyses of geographical variation in relation to ecology may be expected to give much information on the effects of natural selection.

(c) *The study of overlaps.*

Potentially one's most dangerous competitor is, as Darwin pointed out, one's closest relative, who is most likely to have the same needs. Consequently a study of the situation that arises when either two closely related species come to overlap for part of their ranges, or when part of a species escapes from its coexisting relatives by colonizing a distant district, should throw much light on the nature of differences between species. This expectation has already been justified even from the few studies that have been made so far.

Wherever two closely related species overlap, they seem to have distinct ecological preferences, and so avoid competition. A few exceptions have been recorded, but in nearly all of these the observations made were incomplete. Several species of hover-flies, for example, have been found coexisting in exactly the same way; but only the adults were observed, not the larvae, amongst which different food-preferences might be expected.

When two closely related species overlap only partially, it often happens that each one when alone occupies habitats from which it is excluded in the region of overlap by the other species. For example, the snail *Cepaea hortensis* (p. 76) occurs throughout Britain, while *C. nemoralis* extends north only into the southern half of Scotland. Within its range, *C. nemoralis* is the snail commonly found living among old stabilized sand-dunes, but where it is absent in the north, *C. hortensis* is common in this habitat. Similarly, the two European nightingales (p. 137), in the narrow zone of overlap, show distinct-habitat preferences, *L. megarhyncha* restricting itself to drier woods and gardens, *L. luscinia* to more swampy woods. These preferences do not appear to be maintained outside the zone of overlap. In the Solomon Islands there are numerous species of white-eye (*Zosterops*), but most islands have only one, and a few two. Where two species coexist, one restricts itself to mountain forests, the other to lowlands. This does not mean merely that the lower islands only have a single form, since there is only one on some mountainous islands, and usually it is found all over the island (although not always). Similarly, the two species of rosella that have lately overlapped (p. 62) seem to prefer different types of country, *P. adscitus* liking the drier. Although both species may occur together, they seem to separate for breeding. (Of course, where food is locally artificially abundant they may feed together in mixed flocks.)

The beak is the most important organ in most birds for obtaining food, and adaptive variation in its structure is well known. The long thin beaks and tubular tongues of honey-eaters, the short spiky beaks and long tongues of woodpeckers, and the heavy beaks and short tongues of finches eating hard seeds, are obvious adaptations. In his classic study of the 'finches' (the *Geospizinae*) of the Galapagos Islands which so stimulated Darwin, Lack has shown that not only are the different types of beak adaptations to different modes of life (to insect-eating, seed-eating, and so on) but the different sizes of beaks of the same types are adaptations to avoid competition with close relatives. To take only two examples, there are in the ground-finches, *Geospiza*, three species, *G. fuliginosa* with a small but strong beak, *G. fortis* with a large beak, and *G. magnirostris* with a very large one, and these three species normally coexist. There is evidence that *G. magnirostris* can include as part of its normal diet seeds larger than those usually taken by *G. fortis*. Moreover, on two islands, *G. magnirostris* is absent, and on these islands, the beak in *G. fortis* reaches a much greater maximum size than on the other islands. Exactly the same situation occurs in the insectivorous tree finch *Camarhynchus parvulus* on one island where *C. psittacula*, which usually coexists with it and has a much larger beak, is absent. Lack concludes that the variation in beak-size in these groups of species is adaptive, allowing them to coexist without effective competition.

Similarly, in the remarkable case of the two nuthatches discussed above (p. 67), there is a considerable difference in beak-size where they overlap, and moreover, their plumage-patterns are more distinctive there. It is very likely that on the average different foods are taken in the region of overlap, and that specific recognition marks (the importance of which is discussed below) are maintained there. Again, in the two species of fruit-pigeon of the genus *Ptilinopus* which coexist in the Marquesas, it seems that competition is avoided by sharing out the habitat, one being principally found in the mountains, the other in lower country. In correlation with this, little difference in beak-size or body-size has been produced as yet; but it is of the greatest interest to notice that the colour patterns are distinctive. *Pt. mercieri* has a red cap, a pale grey breast, and a bright yellow belly, marked off from the breast by a sharp transverse line, while *Pt. dupetithouarsii* has reduced its cap to white with a narrow yellow border and has dull greenish underparts diversified only by a dull orange blotch on the abdomen. There is far more distinctiveness and general vividness of pattern in these two species, isolated at the extreme edge of the range *Ptilinopus*, than there is in any of their close geographical representatives, all of which tend to degenerate to a plain dull grey or grey-green all over, and are the sole representatives of *Ptilinopus* on their islands, or coexist only (in Fiji and Samoa, p. 139) with strikingly different forms. There can be little doubt that the co-existence of two very closely related forms in the Marquesas has caused the retention and even perhaps intensification of specific recognition marks that have been allowed to degenerate elsewhere. In the Galapagos finches the plumage generally is very dull and inconspicuous, partly for reasons of protection, but also because the very different shapes and sizes of the beaks of the various forms seem to be of considerable value in specific recognition. In the Marquesan *Ptilinopus* such characters have not yet been produced, so plumage differences are maintained.

Two subspecies of the parrot *Trichoglossus flavoviridis* are instructive. *T. f. meyeri* lives in the mountain forests of Celebes, on which island the lowlands are occupied by the closely related and larger form *T. ornatus*, a member of the *haematodus* species-group. *T. f. meyeri* is a small shy, inconspicuously coloured bird, while *T. ornatus*, although protectively coloured on the upper parts (as might be expected in birds of open country) is rather gaudy, noisy and social. On the Sula Isles, just east of Celebes, no member of the *haematodus* species-group occurs, and the only *Trichoglossus* is *T. f. flavoviridis* which is considerably larger than *meyeri*, (almost as large as a small member of the *haematodus* group) and much more vividly coloured, since it has much bright yellow on the head and breast. No good ecological study of this form has been made, but it seems reasonable to conclude that it is occupying a wider niche, more like that of the *haematodus* group, than is its close relative, *T. f. meyeri*.

Such examples as these, and many others, all demonstrate the extreme importance of the relationships between closely allied species, and in general of the relationships between any species and the rest of the fauna, which as Darwin pointed out, are of the greatest importance in developing adaptive characters. They indicate clearly that in general, the selective forces acting on any species will vary greatly with geographical variations in the rest of the fauna. If a species can colonize a district where its close relatives are absent, the new population may be able to invade new niches, abandon some specializations, develop others more suited to its new situation, and in general rapidly acquire profound modifications which cannot be produced where it is stabilized by perpetual pressure from its relatives. Even in the few examples given above, there are strong indications that apparently trivial details of the plumage-pattern may have considerable selective value as recognition-marks or signals for similar functions, and this indication is strengthened by the remarkable demonstrations of the importance of such signals made by Lorenz and Tinbergen. Further studies of the selective effects of overlaps, modelled on Lack's analysis of variation in Darwin's finches, are greatly needed.

(d) *Investigations of selection.*

Many investigations have been made which show clearly the effectiveness of natural selection; in some it has been possible to attempt the difficult task of actually estimating the selection coefficients involved. Almost invariably these have been found to be surprisingly high. Since it is exceedingly difficult to estimate coefficients unless they are high, it might well be felt that those so far obtained are a very biased sample. This may be, but what is of particular importance to us is that nearly all are related to (at first sight) exceedingly trivial and 'non-adaptive' characters, and it is therefore not unreasonable to suppose that those characters whose usefulness is obvious even to the zoologist will be subject to even heavier selection.

For example, da Cunha has investigated polymorphism in the fruit fly *Drosophila polymorpha*. In both males and females of this animal the abdomen is yellow with black interrupted bands, but the extent of the bands is variable, three types, dark (with very broad bands), intermediate, and light being distinguishable in either sex, and found coexisting in natural populations. The genetic basis is very simple and involves two allelomorphs, one in double dose (EE) producing the dark pattern, the other the light (ee). Individuals hybrid for these genes (*heterozygotes*, with constitution Ee) have the intermediate pattern. It was possible to show that if the selective value of the heterozygotes (Ee) is taken as 100, that of the dark individuals

(EE) is fifty-six, and of the light twenty-three. When a heterozygote mates, some of its offspring always belong to one or both of the other classes, or *homozygotes* (EE and ee). Consequently, the homozygotes, although at a disadvantage, will be constantly replenished in the population, and the polymorphism will be stable. Such large differences in selective value between three apparently very trivial pattern-characters is of the greatest interest. One is tempted to ask of what possible importance the possession of one of these patterns rather than anothei can be to the fly. Almost certainly, this is another case where a character probably has no value as such but is merely that one effect out of a whole complex most obvious to man.

Such stable polymorphism as that just described gives much information on selection. Fisher has pointed out that stability cannot be attained unless there is a balance of selective forces. Any valid example of stable polymorphism is evidence as such that the genes concerned are undergoing selection; yet stable polymorphism is often found to affect very 'trivial' characters. Dobzhansky and other workers have been able to show by a series of very elegant experiments not only that in several species of *Drosophila* there is stable polymorphism in respect of various structural rearrangements in the chromosomes, but that the balance of proportions of the different forms is exceedingly sensitive to quite small changes in temperature.

Stable polymorphism is found also in the variations of shell colour and banding pattern in the snail *Cepaea nemoralis* (p. 76). These variations have been frequently described as random or non-adaptive. In fact, here also there must be selection maintaining the stability. It has been shown in addition that several species of bird attack these snails and tend to pick out those which least resemble their surroundings. For example, in dense beechwoods, with their perpetual carpet of red-brown leaf-litter, red and brown shells are common, yellow ones rare, while in very green places yellow shells are abundant. In such places as beechwoods where the carpet hardly changes through the year, coefficients of visual selection alone must be at the very least of the order of 2 per cent, and probably much higher. In a mixed deciduous woodland it was possible to show that selection varied with the season. In early spring, when the floor was brown with leaf-litter, yellow shells were at a considerable disadvantage, being taken preferentially by thrushes, while in early summer when there was much greenery, they were at a distinct advantage to pinks and browns. The delicacy and variability of visual selection in this snail in relation to even slight changes in the background is paralleled by the delicacy and variability of selection by temperature in the chromosomal rearrangements in *Drosophila*.

In the great industrial areas of Britain and Germany several species of moths in different genera now show the phenomenon of industrial melanism. All have rather pale wings, blotched, speckled and banded in such a way that when the animals are at rest on tree-trunks, they are beautifully camouflaged. In and around the industrial areas, their wings are almost, or entirely, black. Ford has shown that these melanic forms are more hardy than the normal. In the moth *Boarmia repandata* feeding of larvae only on alternate days was enough to send up the proportion of melanics among the moths which developed from the survivors, which in these experiments should have been 50 per cent, to about 70 per cent. Even under normal conditions, the proportion of melanics was still 4 per cent greater than expected. Yet in spite of this marked hardiness, the melanics decrease very rapidly in frequency as one moves out of an industrial area. There seems little doubt that in country districts the black forms are at a heavy disadvantage because they are conspicuous and readily found by insectivorous birds. In industrial regions their predators are far less common. Their colour is only slightly disadvantageous, and this is more than counterbalanced by their superior hardiness.

Many such examples as these could be quoted. Onions with deeply coloured bulbs are highly resistant to infection by a fungus, and here again only a single pair of allelomorphs seems to be involved. Slight differences produced by the effects of many genes, or even by temporary changes in complexion are equally subject to selection. Sumner has shown great selective discrimination by fish-eating birds of fish too dark or too light for their background. The fish were made darker or lighter simply by keeping them in the dark (when they quickly became pallid) or in full exposure to light. Sukatschew demonstrated that different genetic strains of dandelion when grown intermixed had very different powers of survival, which varied greatly both according to the degree of overcrowding of the plants, and to climate. Several workers have shown that the application of fumigants on a very large scale in Californian orchards has resulted in the appearance of genetically resistant strains of various insect pests since 1900. For a full discussion of selection the reader should refer to the reviews by Huxley and Dobzhansky listed at the end of this book.

To sum up this section on the production of divergence, it seems thoroughly well established that (a) by far the most important single factor in controlling the variability of a species is natural selection, which in every case where it has been properly investigated seems to act with extraordinary delicacy upon every character examined. (b) The probability that genetical drift is generally important in producing diversity is very slight. (This conclusion is not accepted by some authorities.) (c) Since no two localities on the earth are identical in every feature of soil, climate, flora and fauna, and since even slight differences between localities can produce considerable selective differences in the populations inhabiting them (provided there is not excessive swamping by migration) the production of geographical diversity in populations by local selection is only to be expected.

MAINTENANCE OF SPECIFIC DISTINCTIVENESS

It seems clear, then, that if for any reason a population becomes isolated, and freed from swamping, it will feel the full effects of selection for life in its particular locality, and, as no two localities are exactly alike, it will gradually build up a genetic constitution adapted for life in that locality. The complexities and delicacy of selection and of the interaction of genes and their products within each organism (itself a highly co-ordinated system of unbelievable complexity) are so great that it is in the highest degree improbable that two populations in genetic equilibrium with their environments in two different localities will achieve genetic constitutions which are exactly equivalent in every possible respect. Consequently if individuals from the two populations meet and mate, the hybrids, which will have constitutions made up of parts wrenched almost at random from two different balanced complexes, are most unlikely to put up as good a performance in nature as either of the parent forms;

the further the parents have diverged, the greater will be the disharmonies of the hybrids.

Evidence of such incompatibility has been found experimentally, even within widely separated populations of a single species. The North American frog *Rana pipiens* is very similar to the common European frog *R. temporaria*. Moore has shown that in the eggs of *R. pipiens* there is considerable geographical variation in size, temperature-tolerance, and rate of development, which is clearly adaptive to climate. Regional selection at least for these characters (and certainly for many others) occurs within the species. But Moore also found that if eggs from one district were fertilized by sperms from males in nearby districts, the tadpoles duly hatched, grew, and metamorphosed into young frogs. If eggs and sperms from more distant districts were used, a lower percentage hatched, the tadpoles tended to be slightly malformed, and metamorphosis was less often successful. When crosses were made between frogs from New England and Florida or Texas, the hybrids were exceedingly feeble—in fact, there was effective genetical isolation between these distant populations. Similar examples are known in other organisms. It is evident that if, because of large-scale climatic or geographical changes the extreme populations were to extend their ranges until they met, a ring-species would result, which would produce only sterile hybrids at the overlap, so that the populations there would be genetically separate.

There is little doubt, then, that local natural selection (given time and protection from excessive gene-flow), will build up different harmonious gene-complexes which can come to be incompatible with each other; the populations possessing them, when they meet, will behave as separate species. But this cannot be the whole story, because, as already shown (p. 94) by no means all good species are intersterile. Among ducks, for example, intergeneric hybrids have been produced in captivity which seem completely hardy (although probably not under the much more stringent conditions found in the wild). The barrier between species is by no means always that of hybrid sterility, or failure to produce offspring, but may be due to differences in the courting-behaviour of the two forms, or even to simple mechanical difficulties in mating.

The probable origin of such barriers has been elucidated by

Dobzhansky, who refers to them by the useful term *isolating mechanisms*. Suppose that two forms come into contact in a certain region, and that for any reason whatever (by partial genetical incompatibility at the time of meeting, and by the occurrence of subsequent mutations) the hybrids are slightly less successful than either species within its own territory. Then every mating between individuals of the same species will be more successful than crossings between the species, which will be at a selective disadvantage. Any mutation or other hereditary variation which will in any way facilitate the mating of individuals within the same species, will obviously be at an advantage to others which have no such effect or actually promote hybridization. Consequently, there will be selection for any mechanism which will produce barriers between the two forms.

This hypothesis has already received some experimental verification. The sibling species *Drosophila pseudoobscura* and *D. persimilis* (p. 80), produce hybrids rather readily at rather low temperatures. Koopman kept the two species together at low temperatures for a number of generations and destroyed the hybrids (which in any case were effectively sterile, under the conditions of keeping the animals). He was able to show that after a very few generations, the proportion of hybrids produced had fallen very significantly (from about 35 per cent to less than 5 per cent). Selection had improved the sexual isolation between the two populations with extraordinary rapidity.

It seems that isolating mechanisms will be developed (or at least perfected) when species meet and there is no need to postulate the production in isolation of more than enough incompatibility to make hybrids slightly less viable. Certainly the complexity of some isolating mechanisms, such as courtship dances, shows clearly that they can hardly have been developed in spatial isolation on the off-chance that they will provide genetical isolation from species never yet encountered. But certain conditions are necessary for producing such mechanisms. The hybridization must be neither too extensive nor too restricted.

If hybridization is too extensive, the two populations will simply fuse. Purebred individuals will be so rare, and their production will cease so rapidly that no slight selective advantage can secure their continuance. The result, as in some populations of the golden whistler *Pachycephala pectoralis* p. 138, will be a single hybrid population. In North Africa the house sparrow (*Passer domesticus*) and the willow sparrow (*P. hispaniolensis*) do not interbreed except in central and eastern Algeria, where they hybridize extensively, forming a very variable intermediate. But in southern Algeria, where the available habitats are restricted to a few isolated oases, only the intermediate occurs, and it shows comparatively little variation. In these oases, there are stabilized hybrid populations.

On the other hand, if the overlap is very slight, in proportion to the mobility of the species concerned, very few hybrids will be produced, and they will mate with the parental types far more often than with each other, so that even if barriers against hybridization are produced in some individuals, the combinations of genes producing them are likely to be broken up by matings with the unmodified parental types. In this case, a stabilized hybrid zone, as in the carrion and hoodie crows (p. 95) will be produced. Dobzhansky has suggested that the local variations in width of the hybrid zone between these forms is correlated with the age of the contacts in different places, selections for isolating mechanisms having reduced the zone where contact has been established longest; but Mayr points out that this explanation does not fit the facts. The width seems to vary according to local ecological conditions. This implies that the strength of selection against the hybrids varies according to environmental conditions, which might well be expected. The stability of the zone must therefore be due on the one hand to the continual production of hybrids, since isolating mechanisms, if developed, are continually broken down, and on the other to the selective disadvantage of the hybrids relative to the parental forms in different types of country.

A very convenient simple classification of isolating mechanisms is due to Mayr.

(1) *Ecological.* Mating does not take place, because individuals in breeding condition do not come into sufficiently close proximity, either because they are spatially isolated by unsuitable terrain, or because their breeding seasons are separate. (The first of these alternatives may be reversible by simple geographical or climatic change. The second, and all other mechanisms are not simply reversible.)

(2) *Ethological.* Mating is preceded by the mutual production in correct sequence of appropriate stimuli, which must be appropriately answered, before copulation will be allowed. The highly specific recognition markings and courtship behaviour of many birds and fishes, the complex courting 'dances' of slugs and some butterflies, the special calls of certain birds, scents of different mammals, and butterflies, the specific rhythm of wing-vibration in courting *Drosophila*, are only a very few examples of structural and behavioural features that enable individual animals to recognize and mate with those individuals that can be expected to produce normal and fertile progeny. All such ethological mechanisms are isolating mechanisms of great efficiency, since they prevent wastage of the gametes.

(3) *Mechanical.* Mating is attempted but is prevented or nullified by mechanical difficulties. Not many examples of this are known as yet, but in certain species of *Drosophila* it seems that wrongly mated individuals may be unable to disengage themselves after mating. Mechanical difficulties are probably important in preventing mating between some large and small species of snails and earthworms.

(4) *Reduction of mating success.* In all animals that merely eject the gametes, which fuse in the surrounding water, and in all others where either no other mechanism is present, or breakdowns have occurred, either wrong gametes may be unable to fuse, or they may produce sterile eggs, or the juveniles may fail to reach maturity, or if they do, they may (as in mules) be sterile or at some lesser but still important selective disadvantage. Such physiological barriers are in one way the least efficient, since in general they only come into action after gametes have been released and so wasted; but in many organisms they may be the only possible barriers, and in all they are the final safeguard against occasional breakdowns of other sorts.

To sum up this section, it seems evident that genetical divergence and subsequent selection, perfecting isolating mechanisms, can account both for the maintenance of specific barriers, and for the elaboration of some of the mechanisms used in maintaining them.

ECOLOGICAL COMPATIBILITY

Even if two forms meet along a line of contact and are able to preserve perfect genetical isolation, they may not be able to coexist in more than a genetical sense (p. 94) unless they can achieve ecological compatibility. If they are adapted to different habitats which are sharply marked off from each other geographically, then one species will be found only in one habitat, the other in the other, and there will be local replacement, as in the grouse and ptarmigan (p. 89). If the habitats occur in compact masses (like a large block of rain forest bordering a stretch of savanna) there will be geographical replacement in every sense. Mayr has pointed out that some good species which replace each other completely within their compact ranges have been mistaken for subspecies because they are very alike and apparently geographically representative, but a study of the line of contiguity shows no intergradation.

The principle that a homogeneous environment cannot support two very similar species was deduced and experimentally verified by Gause in 1934. It is in the highest degree improbable that the two forms would be identical, consequently the better-adapted one will eliminate the other. This conclusion is completely confirmed by observation on closely related species in the wild. Invariably, they are found to differ in their ecological requirements in some respect. Either they take the same sort of food but live in different climates or vegetation, or if they coexist, they take rather or very different food; and other necessities besides food are similarly shared out. In the very few cases where closely related species have been considered to live in exactly the same way, insufficient evidence has been obtained.

Broadly speaking there are only three ways in which new populations can become established. The colonizing species must extinguish resident ones, or it must share the available resources with them, or it must find unoccupied territory. All three processes are of great importance in evolution, but the first and third solve the problem of coexistence merely by avoiding it. A burst of colonization may be brought about by accident (distribution by storm, or man), by overcrowding due to population increases in exceptionally favourable seasons, or by genetical changes allowing increases in population or producing better tolerance to particular conditions. This is a subject of which very much remains to be learnt.

A study of the overlaps (mentioned above, p. 153) of *Cepaea hortensis* and *nemoralis*, the Geospizinae, and many others, shows that where two similar forms coexist, they may share out habitats which either is able to occupy completely when by itself. In the areas of coexistence, each is forced by the presence of the other to specialize. Now it is clear from what has been said above about the effectiveness of natural selection, that the more completely the ranges of two forms are coextensive, and the longer they are in contact, the stronger will selection be for specialization in those habits which are most likely to cause conflict between the two forms. There will be selection for anything promoting the production of ecological isolating mechanisms, just as there will be for the reduction of gene-flow in the development of species-barriers, but with this important difference that whereas two coexisting species can readily become genetically separate, it is very doubtful whether they can ever cease to have some ecological influence on each other as long as they coexist.

Further, the greater the difference that can be produced between two populations in isolation, the more readily they will coexist on meeting. The two very similar species of nightingale in Europe (p. 153) and other such pairs are too similar to be able to coexist except in a very narrow region of overlap, while those species such as the tree-creepers (p. 137) which differ rather more in their requirements are able to overlap extensively in their geographical ranges. The three species of *Geospiza* (p. 154) on the other hand, have specialized sufficiently to be able to coexist completely.

In the region for which it is best adapted, a species is (naturally) comparatively abundant, whereas in unfavourable regions it may be able to exist only in a few exceptionally favourable situations. In consequence it is often found that populations near the edge of the range of a species are ecologically much more restricted than those near the centre. In the damp, cloudy, mild oceanic climate of Britain, many species that are widespread on the continent of Europe are very restricted in range, and occur in just those regions that might be expected to give them the best approximation to more continental conditions. Several species of snail are confined in Britain to the well-drained, easily warmed chalk downs, or to them and some of the purest limestones in the Cotswolds and Pennines, while on the Continent they are not especially characteristic of extremely calcareous districts. The extreme limitation in Britain of the Glanville Fritillary *Melitaea cinxia* to the south shore of the Isle of Wight is probably due to its need for considerable sunlight, since as Ford has pointed out, the caterpillars are found on a great variety of soils, and this region is the sunniest in Britain. The caterpillars feed on the leaves of two species of plantain (*Plantago maritima* and *P. lanceolatus*) which are abundant there. Elsewhere in Europe this butterfly is not confined to the coast, and the caterpillar feeds on a greater variety of herbs. It appears that its extreme localization in Britain has brought about a restriction in food also. If the British populations could increase in numbers, and invade the Continent, it is quite possible that they might displace the Continental forms in cloudy regions where such plantains are common.

Another instance of specialization at the edge of the range is given by the very distinct British subspecies of the common swallowtail butterfly, *Papilio machaon britannicus* which is confined to the Fens, where its caterpillar feeds on milk parsley (*Peucedanum palustre*). On the Continent, this butterfly is not at all confined to fenland, and far more food plants are known. But recently specimens of the French subspecies *P. m. gorganus*, have appeared frequently, and caterpillars have been found, in the south-eastern corner of England, which has a rather continental climate. It is very likely that a reinvasion of Britain is taking place. If so, two subspecies, one highly specialized in diet and confined to fenland, the other eating various plants, (especially wild and cultivated carrots) and preferring dry situations, will be found quite close to each other, and might be considered ecological replacements.

Many other examples of specialization in peripheral populations could be given. In each, selection can produce a considerable divergence from the normal in ecology, so that, should conditions change, these populations may well be enabled to colonize previously uninhabitable regions, or to coexist with other populations without violent competition.

The predisposing influence of local selection to partial ecological compatibility is well seen in some Australian birds. The south-western refuge (p. 136) became drier than the south-eastern, and the birds that survived in it were those that adapted themselves to the local conditions. In both the parrot *Barnardius* and the tree-runners *Neositta*, and in some other genera, the south-western forms were able to spread widely when the climate became slightly less arid, and have advanced across the head of the Great Australian Bight into the south-east, where they occupy drier ground than their south-eastern representatives. In the very centre of Australia, in and near the Macdonnel Range, is a region where several species of plants survive as relics, separated from their nearest relatives by huge areas of arid country. This refuge became too dry for most birds, but a very few managed to survive in it, adapting themselves in course of time to well-nigh desert conditions and spreading out from it over all but the completely barren areas of the interior. The little night parrakeet, *Neophema bourkii*, and the Princess Alexandra parrakeet, *Polytelis alexandrae*, each the only one of its genus to inhabit such areas, are almost certainly a product of slow selective transformation in this failing refuge. Their habitats are consequently very different from those of their nearest relatives.

Similarly, some species of the big nutmeg-pigeons (*Ducula*) and of the small fruit-pigeons (*Ptilinopus*), some subspecies of the golden whistler, and many other forms, are birds of small islands and coastal woods in the New Guinea region. They could have become restricted to these habitats either by colonizing unoccupied small islands, becoming adapted to their maritime flora and climate, and spreading from them successfully, or by being restricted to coastal woods on the large islands through competition with other species, specializing, and so facilitating the invasion of small islands, or by being marooned on a sinking archipelago of large islands, and surviving by adaptation. Probably all these processes were involved. A close study of geographical and ecological variation in many groups in the same region will bring to light many more such examples as these.

To sum up this section, we can say that forms with sufficiently different ecological needs can invade each other's ranges and coexist ecologically. Geographical variation of ecological preference is widespread, and particularly noticeable in peripheral populations. It is produced by local selection, whether for climatic or similar reasons, or because of the presence or absence of different competitors. Any character that will promote ecological compatibility will be at an advantage where two similar species meet, and both will be modified by their meeting. The modification may be at first merely a simple change in behaviour but can produce profound adaptations in all parts of the organism.

To sum up this chapter, it appears to be firmly established that spatial isolation is of the first importance in producing new species. Although the characters by which geographical races differ may often seem trivial and non-adaptive, they are probably always either directly adaptive as such or by-products of some other character or process that is adaptive. Local selection will produce genetical and ecological divergence of populations which if they meet may be able to develop very efficient isolating mechanisms to keep themselves genetically separated, and ecological specializations to reduce competition and promote ecological compatibility. Within the theory of geographical speciation many particular problems remain to be solved. Indeed, here is one of the most promising fields for research in evolutionary zoology. But the general theory seems well established.

It is obvious that the same processes that produce divergence in two separate populations produce also the transformation of either population considered separately. With very few exceptions, palaeontologists agree that these processes are the basis of transformation in time, within a gens. That is why it was said above (p. 135) that fundamentally there is no difference between anagenesis and cladogenesis. One may add that while the relative importance of selective and non-selective processes remains to some extent a matter for controversy, every addition to our knowledge of genes in natural populations serves only to emphasize again and again the power and scope of natural selection.

CHAPTER IX

SYMPATRIC SPECIATION

WHILE no one now denies the great importance of geographical speciation, many have felt that there is little evidence for it in certain groups. Darwin wavered considerably in his opinions and many passages in the 'Origin' are ambiguous or can be taken to mean speciation within a single population. He mentioned particularly (p. 131, above) the difficulty of species gradually replacing each other in different parts of the same continent, and was inclined to believe that local selection could cause parts of a widespread population to diverge so far that they could become separate species.

SEMISYMPATRIC SPECIATION

Exactly how far parts of populations can diverge obviously depends on the strength and periods of action of local selection and the mobility of the species. It is not impossible that if a species comes to occupy two adjacent but very different habitats, sharply marked off from each other, the individuals in each will undergo heavy selection, and the hybrids between them may be at a high disadvantage in both pure habitats, being adapted to neither. This appears to be the case in the carrion and hoodie crows (p. 95) and is a situation rather well known in plants. In a small fen near Oxford is found a very localized population of the orchid *Orchis traunsteineri*. On the dry ground around it occurs *O. fuchsii*. The two species flower at the same time. At the very edge of the fen, where they approach within a few feet of each other is a swarm of hybrids showing every combination of parental characters. Orchid seed seems to be rather easily distributed, and the population of *O. traunsteineri* is small, yet in the middle of the fen, it shows few signs of hybridity. Obviously there must be very heavy selection reducing gene-flow between the two populations. It is highly significant that the hybrids appear to maintain themselves only in the very narrow strip intermediate geographically and ecologically between the two habitats. Muller has described a very similar situation on a far larger scale in several species of North American oaks, all normally with very strict localization on particular types of soil. Hybrids occur only where two soil-types are adjacent, and are plentiful only where they intergrade in nature; where the change from one soil-type to the next is sharp, hybrids are very few. Since oaks are wind-pollinated and it can be shown from the existence of hybrids that the species concerned have no great genetical incompatibility, very heavy local selection is the only possible explanation for the rarity of the hybrids. It is noteworthy that those species of these oaks that coexist over large areas hybridize only rarely, if at all, and that the geographical approach of some of the hybridizing species is believed to be very recent, so that isolating mechanisms may not yet have been developed. A similar situation is known in *Drosophila*. Of three very closely related species, *D. guaru*, *D. guarani* and *D. subbadia*, the first two are Brazilian sympatric forms and do not interbreed, but both interbreed rather freely (in the laboratory) with the third which is a Mexican form, allopatric in the wild to both of them.

The appearance again and again in the same sort of habitat of a genetically distinct form interbreeding more or less freely with a different form found in adjacent habitats, is very common in plants. Such genetically determined forms are known as *ecotypes* and there seems to be every gradation between them and good species excluding each other in different habitats but with much contiguity and no hybridization. In some groups of ecotypes, the hybridization seen in intermediate areas has every appearance of being secondary (p. 88) as in the oaks and orchids quoted above, while in others it seems to be primary, and the ecotypes to be parts of a single population modified by heavy local selection. Probably most geographical subspecies should be considered ecotypes with a single area of occurrence.

As Muller points out, there is no doubt that the distinctness of the hybridizing species of oaks investigated by him is maintained by very heavy selection. But it is far from certain that they arose by local differentiation of a single widespread population, not by change in spatially isolated forms. Regional differentiation leading to specific status in the face of gene-flow may be called *semisympatric speciation* since it involves ecological allopatry and genetical contiguity. How far it is important in nature is unknown. If selection is strong enough and intermediate individuals at a sufficient disadvantage, it is perhaps possible that isolating mechanisms might develop, with the consequent removal of the hybrid zone. This is a subject almost undeveloped at the present day. Mayr, who is the principal authority on sympatric speciation, points out that in many cases in animals hybrid zones are due to recent overlaps, and there is no reason to suppose that the parental forms were not produced by normal geographical speciation. This is often but perhaps not invariably the case for plants as well.

BIOLOGICAL RACES

In several groups of animals, the different forms are almost or (apparently) completely indistinguishable morphologically and can be separated only by their ecological requirements, which, however, may be extremely different and clear-cut. Such forms are often called *biological races*, and are particularly evident amongst parasitic groups and other groups with very specialized food-requirements. The caterpillars of some moths, for example, are monophagous (feeding on only one species of plant), or oligophagous (feeding on only a few), while others can exist on a wide variety of foods and are termed polyphagous. Several genera of small moths contain a large number of very similar species, all overlapping and all monophagous but attacking different plants. One gets the impression that ecological specialization has been far more important than any other process in their evolution. Similarly, among the parasitic protozoa the form which produces the disease dourine in horses is morphologically identical with that producing surra; but the first is a venereal disease, the second affects the blood. These two protozoans, medically distinct and without any intermediates are actually given specific rank as *Trypanosoma equiperdum* (producing dourine), and *T. evansi*. Other 'species' in the same genus cannot attack the same host. Thus *T. brucei* which causes the disease nagana in various hoofed animals cannot be made to infect man, whereas other species can. Even within these 'species' there may be distinctions. It seems that surra is caused in horses, cattle and camels by distinct strains of *T. evansi*. In the malaria parasite *Plasmodium*, found in both man and the chimpanzee, different strains differ in virulence, some causing distressing symptoms only in one host-species, others in the other. In many parasites some strains are much more virulent than others in the same host-species.

Among oligophagous moths it has sometimes been found that two or more biological races, each attacking only a single species of plant, can be distinguished within a single 'species'. Thorpe has described how caterpillars of two such races kept strictly to host-plants even when these were growing so closely together that their branches intermingled. One race could be transferred to the food-plant of the other but only with difficulty, and when a choice was allowed, preferred its usual food. Monophagous species can sometimes be transferred successfully to new plants, but usually there is a very high mortality at first, and much selection has to take place before the transfer is established.

The occurrence of such biological races, and the coexistence of large numbers of very closely related species ('species-flocks') differing principally in ecology, have led many authors to suggest that sympatric speciation, by change of habitat-preference, may be much more important in certain groups than spatial speciation. Sympatric speciation could come about either by slow change of a particular variety, or instantaneously by production of individuals fertile among themselves but unable to mate with their parents. One method of sympatric speciation is indeed known, and is of importance in many genera of plants. It is discussed briefly below. Apart from it, no certain cases of sympatric speciation are known, and there are important objections to the theories of its mechanism that have been proposed. These objections have been especially urged by Mayr.

DIFFICULTIES OF SYMPATRIC SPECIATION

In the first place, the term 'biological race' refers to a very heterogeneous assortment of forms, including not only strains of parasites and oligophagous forms, but subspecies, ecotypes, and even sibling species such as *Drosophila pseudoobscura* and *D. persimilis*, and the species of the *Anopheles maculipennis* complex (p. 78). All that these have in common is that their ecological peculiarities are more striking than their morphological ones. At least the ecotypes, subspecies and sibling species should be excluded from the category, or clearly distinguished within it.

Secondly, all the examples of biological races that have been put forward as inexplicable by geographical speciation are in fact so insufficiently analysed that no conclusion either way can be reached from the available data. For example, the codling moth *Carpocapsa pomonella*, which normally attacks apples and pears, was reported in 1909 to be doing damage to walnut crops in California, and has since become on occasion a serious pest. It has infested apples and pears in the same region at least since 1880. Is this sympatric speciation? As Mayr points out, the example has never been properly analysed. In the chief walnut district apples and pears are infrequent. It is perfectly possible that we have here only geographical speciation in a very sedentary species. But how the insect first came to attack walnuts is not known. Possibly a large number of adults were carried into the walnut district by accident, a few just managed to maintain themselves, and an adapted population was built up. Or constant dispersal from overpopulated regions might have had the same effect provided selective forces in the new habitats were strong enough to kill off all but a few individuals, to offset swamping. In this case distinction would be maintained by violent ecotypical selection, as in the oaks mentioned above. There is certainly not the slightest reason to suppose that a few individuals whose genetical constitutions were such that they had some chance of surviving on walnuts were mysteriously activated to go out of the apple-growing districts where they were produced and to look for that very tree on which they had best chances of surviving.

It is known that this moth does show considerable facility in producing local races. Armstrong has described a Canadian population living in a large old pear orchard more or less isolated from other large orchards. The moths emerge on the average a fortnight later than in other populations, but the delay means that their larvae are ready to penetrate the pears just as they become soft. But only its isolation allows this population to persist as such, since there is enough overlap in the normal and abnormal times of emergence to allow extensive interbreeding and consequent swamping of local selective effects if normal populations were near enough.

Thirdly, as shown above for the codling moth, none of the examples put forward as due to sympatric speciation are inconsistent with geographical (or at least micro-geographical) speciation. Some large species-flocks have been quoted as evidence here. In Lake Nyassa, for example, there are no less than 174 endemic species of Cichlid fishes, and fifty-eight in Lake Victoria. Over 300 species of crustaceans belonging to the one family Gammaridae are known from the huge and ancient Lake Baikal. Similar situations are known in other large lakes. It seems impossible to deny that many species must have arisen within a single lake. But further analysis of the freshwater fauna of Central Asia has shown that many species believed to be endemic to Lake Baikal are in fact widespread. Other lakes also are served by many different river-systems which could provide isolation. Mayr comments "The fact that the species of these swarms are now sympatric and that they live according to the Gause principle in different ecological niches in order to minimize competition has led previous authors astray. Rensch . . . has indicated the right solution. It is that these species have come into contact only after they had evolved and after they had acquired their ecological differences." While multiple colonization is certainly a very important factor in increasing the numbers of species in lakes, as on islands, it should be added that most authors have taken far too anthropomorphic a view of lakes. One appreciates directly the floral and geographical features of a large island and would never regard it as ecologically a single uniform locality. Looking at the uniform water-surface of a lake or the sea, many people forget the ecological diversity beneath. There is certainly no need to believe that 174 invasions were necessary to produce the 174 endemic Cichlids of Lake Nyassa. To those species that inhabit only shallow water with a muddy bottom, the depths of the lake and the rocky parts of the shoreline are perfectly effective spatial barriers. Spatially isolated populations can exist and diverge as well in great lakes as on great islands.

Again, Petersen has pointed out that some monophagous and oligophagous genera of small moths are far richer in species than polyphagous ones. This fact is often quoted as evidence for the isolating effects of ecological speciation; but such an interpretation of it cannot be maintained. On the one hand it is far more likely that a large number of very similar species with different and restricted food-requirements can coexist, than a large number all taking a variety of foods, which are almost certain to come into serious competition with each other. And, on the other hand, if one were to take the map of Europe, for example, and plot the *exact* distribution first of several species of plants taken singly and then of several lumped together, it is obvious that if restricted and highly discontinuous ranges are found they will be in the first set of maps, not the second. Widespread species with great ecological tolerance are far less likely to produce isolated populations and undergo extensive geographical speciation than species whose rather specialized requirements limit them to one particular type of habitat.

Sibling species have also been quoted as evidence for ecological speciation. But as shown above (p. 75) there is every intermediate between extreme siblings and 'normal' species. Siblings are not a separate class. There is no evidence that sibling species differ genetically in fewer respects than others, and at least in *Drosophila*, mutation rates and variation generally seem as high as in other forms. Obviously, selection must be stabilizing their characters. It is easy to see how siblings can arise by normal geographical speciation. The production of the same character by very different genetical arrangements is well known. Selection for physiological characters can change the genetical constitution of a form although its morphological characters are held constant by selection. If two forms can avoid competition either by a simple increase in size, or by physiological differentiation, complex morphological change may be unnecessary. It is known that the larvae of some species of *Drosophila* are better nourished by some yeasts than by others. The form of body required to move and breathe in a semi-liquid environment, to obtain large quantities of yeast-cells from it, and to escape from the glutinous mess in order to pupate may well be almost standardized within a species-group of *Drosophila*, while a slight difference in the physiology of the larvae may be sufficient to ensure that they require rather different diets and are ecologically compatible. Similarly the use of different modes of behaviour rather than gross morphological recognition-characters may allow sufficient sexual isolation. Much of the astonishment caused by the similarity of sibling species arises from an over-emphasis on morphology and corresponding neglect of physiology. If there were two species of man, both oligophagous, but with different diets, this disproportionate emphasis on morphology would probably never have arisen.

But, fourthly, there is one very serious genetical difficulty to be overcome by any proponent of gradual sympatric speciation. If a new form is indeed produced, either as a single individual or a small population, then if it is truly sympatric with the normal form, it will be swamped by interbreeding. In all organisms at some stage of the life-history there is a dispersal phase. In many marine animals the eggs or larvae are distributed by currents. In most terrestrial animals the adult or late juvenile is the highly motile phase. It is true that many species show considerable active selection of the habitat for which they are adapted. Birds that are summer visitors to Britain may travel hundreds of miles to settle down in the one patch of suitable habitat in a whole county. But this is no help. If all the birds bred in one patch return to the same patch to breed, they are a geographically separate population. If, as is normal, they do not, then there is gene-flow between the populations in the various separate patches of that particular habitat. In the first case, speciation (if any) is geographical, in the second it will be prevented by swamping. A way out of this difficulty is to postulate that genetically aberrant individuals arise within a population which have not only changed their ecological preferences abruptly and in the same direction but also have a far stronger tendency to mate together than with normal individuals. Such a situation seems most unlikely, and is unsupported by evidence.

One may summarize the objections to sympatric speciation by saying that the hypothesis is unsupported by any good evidence, that it is unnecessary since spatial speciation could account for all the examples brought forward in support of it, that it gets over the difficulty of swamping only by making a series of improbable assumptions, and that spatial speciation avoids this difficulty without requiring such assumptions. There is no doubt that spatial speciation is a well-supported theory of demonstrably wide application, while sympatric speciation is very dubious.

However, one must be cautious. It may be that as morphological differences are far easier to investigate than physiological ones, the present emphasis on geographical speciation may turn out to be for some groups as excessive as the present practical emphasis on morphological species as against sibling species (which require considerable investigation before they are recognized). Two situations especially are almost uninvestigated.

First, in those species that are parthenogenetic for many generations, a variant might arise that could perpetuate itself during the season of parthenogenesis, so that a large local population of it might appear which might then inbreed. But even so, it is difficult to see how without some spatial isolation the population can maintain itself when sexual reproduction sets in.

A more likely field for investigation is that of parasitic strains and highly specialized feeders generally. It was mentioned above (p. 158) that in onions the genetic basis of resistance to a certain fungus was extremely simple, involving only a pair of allelomorphs. The defence-mechanism is the presence of substances poisonous to the fungus in the outer cells of the onion. It is at least conceivable that a similarly slight genetic alteration in the parasite might enable it to overcome this barrier.

The red scale insect *Aonidiella aurantii* is one of those orchard pests in California that has developed resistance to fumigation with hydrogen cyanide since 1900. It has been shown that resistance as against non-resistance has a very simple genetical basis. It has been suggested that the physiological basis is the ability to keep the breathing apertures closed for a sufficient length of time, but this is not yet certain.

It is possible, then, that in some cases at least, the resistance of hosts to parasites or specialized feeders may be overcome by comparatively simple genetical changes. Consider a blood-parasite infecting one species of antelope (say) and carried by a fly which bites several species. Then if one mutant individual can survive injection into a different species, it may build up a huge population by asexual reproduction in a very short time, change to sexual reproduction, undergo improvement by natural selection, and spread very rapidly indeed through the newly infected species. The parent form will not be able to compete with it in the new species. It may well be able to return to the old host-species, but will be genetically swamped by reproduction with the parent form, while in the new host-species it may flourish, undergoing continual selective improvements in the meantime. The two strains of the parasite would be kept apart by heavy habitat selection, as in the oaks described above, but since the two habitats would consist of highly mobile volumes each able to move around the others (and each separated from the others by an environment utterly unsuitable for the parasites), divergence must be called sympatric. The problem is whether a form differing from the normal only by a comparatively simple genetical change can penetrate a highly efficient barrier into a new environment and there undergo sufficient selection to produce some ecological incompatibility with the old environment and genetical incompatibility with the parent stock. It is in essence the same as the problem of semi-sympatric speciation.

POLYPLOIDY

Instantaneous speciation raises even more difficulties. It requires the sudden production of individuals that are not crossable with their parents, but crossable and fertile with each other. But the differences between species, and the genetical bases of crossability are complex and hardly to be produced at a single bound. In some flowering plants, however, where fertility is controlled to a very great extent by the interactions between the pollen grain and the stigma it happens to be deposited on, genetically simple mechanisms bringing about incompatibility between plants and their own pollen are well known. Such devices prevent excessive inbreeding, and consequent loss of genetical variability. The sudden production of a form incompatible with all pollen but its own has not been observed. Various plants, it is true, do practise *cleistogamy*. In them the flower buds never open, and fertilization is effectively internal, foreign pollen being unable to enter the closed bud. But obligate self fertilizers, which are probably uncommon, while not agamospecies, should be classed with them, since in both, each individual has only a single parent. The taxonomic consequences of this are discussed above (p. 98).

One genetical mechanism of importance is known which can produce instantaneous speciation. This is polyploidy. Generally speaking, each species has in the cells of each individual, a characteristic number of chromosomes, those bodies that appear in some sense to bear the genes or factors controlling inheritance. In sexual reproduction a special cell (the gamete), bearing a set of chromosomes derived from one parent fuses with another similarly equipped and derived from the other parent. The resulting cell (the fertilized egg in animals) obviously contains two sets of chromosomes, and is termed *diploid*. Consequently when the individual derived from it requires gametes, it must produce them by a special cell-division producing *haploid* cells with only a single set of chromosomes. Occasionally at some stage in this cycle an abnormal division may occur. In a dividing diploid cell the chromosomes all divide, to produce the full equipment for each daughter cell. But rarely, the rest of the cell may not divide, and a cell with twice the usual set of chromosomes may be produced. Or diploid instead of haploid gametes may occur, by a similar failure. If the diploid number is $2n$, then cells with $3n$ are called triploid, $4n$ tetraploid, $5n$ pentaploid, and so on. In general such cells, and whole individuals derived from them are called *polyploids*.

It is an essential feature of sexual reproduction that before the production of the haploid gametes, the chromosomes of the cells directly or indirectly parental to the gametes come together in homologous pairs and exchange portions of their substance. This ensures that all the particular genes carried by one chromosome are not always inherited as a single block but some at different times may be exchanged into chromosomes which may be carrying at corresponding places alternatives or allelomorphs. A high degree of variability is thus maintained.

Suppose that a normal haploid gamete with n chromosomes should fuse with an abnormal diploid one with $2n$. The result is a triploid ($3n$). If it attains maturity, it will have the greatest difficulty in producing gametes, since the chromosomes of one of its three sets will have nothing to pair with before the reduction division, and will be randomly distributed to the daughter cells. Abnormalities involving whole chromosomes normally cause serious disturbances in the cells carrying them. Consequently, triploids are normally almost entirely sterile, although they can, of course, reproduce vegetatively.

If a tetraploid cell gives rise either to gametes (which will be diploid) or by vegetative reproduction to separate individuals, then tetraploid individuals will be produced. If these cross with the parent stock, sterile triploids will be produced. In fact, polyploids are sterile to the parental diploid. For various reasons they are often not as fertile to each other sexually as diploids, but may be as successful vegetatively. Very many polyploids are in fact known in many groups of plants, and there is no doubt that their production has been an important source of species.

Polyploidy opens up a possibility of speciation by hybridization. If a hybrid between two species is produced, all its cells will have one haploid set of chromosomes derived from one parental species plus one derived from the other, and these will often be too different to pair successfully, so that the plant will be sterile. Its chromosomal constitution can be represented as A+B. If doubling occurs, the new cells will have A+A+B+B, and normal pairing, A with A and B with B, is possible. Such hybrid forms, derived from two species, are called *allotetraploids*. Their formation in plants in the wild and under cultivation has been detected, and in at least one case, a good species, found in the wild, has been synthesized from its parent forms.

Polyploidy seems to be uncommon in animals except among wholly parthenogenetic forms. It is not infrequent, for example, among lumbricid earthworms, where species with high chromosome numbers, probably at least octoploid, have been found. As the inheritance of maleness and femaleness appears to be controlled in most animals by an interaction between certain special chromosomes and the rest, it has been suggested that multiplying of whole sets will break down this balance and produce very ineffective individuals. Parthenogenetic forms will not suffer from such a restriction. But polyploids have been found in at least one species of plant in which sex chromosomes seem to occur, and there is evidence of their production in some fishes, at least. However, it is certainly true that far fewer have been found in sexually reproducing animals than in plants. They are of the greatest interest to the zoologist since they point to a diversity of possible genetical systems allowing very different possible modes of speciation, and to the problem of why they are not abundant in most of the animals investigated so far.

To sum up, it can be said that while spatial (geographical) speciation is well authenticated in animals, sympatric speciation is not. Moreover, there are several serious difficulties that seem to vitiate most theories of truly sympatric speciation. But in plants, and perhaps in some sexually reproducing animals, one genetical mechanism (polyploidy) is known which does produce instantaneous sympatric speciation. While it is undeniable that spatial speciation is common and widespread in animals, the possibility that sympatric speciation may occur in a few groups cannot be excluded.

CONCLUSION

IN the last chapter of the *Origin of Species* Darwin summed up his views on the species question as follows:

> When the views entertained in this volume on the origin of species, or when analogous views are generally admitted, we can dimly foresee that there will be a considerable revolution in natural history. Systematists will be able to pursue their labours as at present; but they will not be incessantly haunted by the shadowy doubt whether this or that form be in essence a species. This I feel sure, and I speak after experience, will be no slight relief . . . Systematists will have only to decide (not that this will be easy) whether any form be sufficiently constant and distinct from other forms to be capable of definition; and if definable, whether the differences be sufficiently important to deserve a specific name . . . Hereafter we shall be compelled to acknowledge that the only distinction between species and well-marked varieties is, that the latter are known, or believed, to be connected at the present day by intermediate gradations, whereas species were formerly thus connected . . . In short, we shall have to treat species in the same manner as those naturalists treat genera, who admit that genera are merely artificial combinations made for convenience. This may not be a cheering prospect; but we shall at least be freed from the vain search for the undiscovered and undiscoverable essence of the term species.

How nearly right Darwin was can readily be appreciated by considering that of the four meanings of 'species' distinguished at the present day, namely, morphospecies, agamospecies, palaeospecies and biospecies, his words are completely applicable in theory to two and in practice to a third. With respect to the biospecies, he was predisposed to error by his insistence on the truth that specific rank is only one of a series of ranks passed through by evolving forms. And this attitude was reinforced because some naturalists had asserted "that species, when intercrossed, have been specially endowed with the quality of sterility, in order to prevent the confusion of all organic forms." ('Origin', Chapter VIII.) It was therefore his business to show (correctly) that there is every degree of infertility between both forms ranked as varieties and species, not a clear-cut separation of varieties and species. In doing so, it seems that he lost sight of the significance of intersterility (or better, of lack of crossability) in preventing the confusion of all organic forms. The cause of his error was only over-appreciation of certain truths.

At the present day it is universally recognized, as a result of Darwin's work, that the species is indeed only one state in the process of evolution. But it must now be added that in perhaps the most important of its meanings, it has a unique significance in being that stage of the evolutionary process at which evolving populations become genetically independent of one another and found new phyletic lines. All ranks above the species refer only to particular groups of these lines, all ranks below only to parts of a single line. The biospecies is a single phyletic line considered during a time-quantum and as such is less artificial, subjective, or arbitrary than any other natural rank.

SUGGESTIONS FOR FURTHER READING

BOOKS

(a) Taxonomic practice.

CALMAN, W. T., 1949. *The classification of animals*. Methuen, London.

FERRIS, G. F., 1928. *The principles of systematic entomology*. Stanford University Press, California, and Oxford University Press, London.

SCHENK, E. T. and MCMASTERS, J. H., 1936. *Procedure in Taxonomy*. Stanford University Press, California, and Oxford University Press, London.

Lectures on the development of taxonomy. 1948-49. Linnean Society of London, 1950.

Lectures on the practice of botanical and zoological classification. 1949-50. Linnean Society of London, 1951.

(b) Evolutionary implications.

MAYR, E., 1942. *Systematics and the origin of species*. Columbia University Press, New York.

DOBZHANSKY, T., 1951. *Genetics and the origin of species*, 3rd edition. Columbia University Press, New York.

LACK, D., 1947. *Darwin's finches*. Cambridge University Press.

JEPSEN, G. L., SIMPSON, G. G., and MAYR, E. (editors), 1949. *Genetics, paleontology and evolution*. Princeton University Press.

HUXLEY, J. S. (editor), 1940. *The new systematics*. Oxford University Press.

PAPERS

Important papers will be found especially in the journal *Evolution* (Society for the Study of Evolution, Lancaster, Pa., U.S.A.), but also in many others. The books and papers in this list have very full references to the literature.

MAYR, E., 1947. Ecological factors in speciation. *Evolution*, 1, 263-288.

MAYR, E., 1948. The bearing of the new systematics on genetical problems. The nature of species. *Advances in Genetics*, 11, 205-237.

MAYR, E., 1949. Speciation and selection. *Proceedings of the American Philosophical Society*, 93, 514-519.

SIMPSON, G. G., 1951. The species concept. *Evolution*, 5, 285-298.

COLLATERAL READING

FORD, E. B., 1934. *Mendelism and evolution*, 2nd edition. Methuen, London.

WADDINGTON, C. H., 1939. *An introduction to modern genetics*. Allen and Unwin, London.

GEORGE, T. N., 1951. *Evolution in outline*. Thrift Books, London.

SIMPSON, G. G., 1951. *The meaning of evolution*. New American Library, New York.

HUXLEY, J. S., 1942. *Evolution: the modern synthesis*. Allen and Unwin, London.

SIMPSON, G. G., 1944. *Tempo and mode in evolution*. Columbia University Press, New York.

INDEX

Aberratio, 47
Acanthiza in Tasmania, 140
Adaptive radiation, 22; in creodonts, 23
Agamospecies, 98; definition, 103; evolutionary potentialities, 103; relationships, 122
Allelomorph, 144
Allen, 151
Allolobophora, good species, 75
Allopatric forms, 73, 91, 93
Allotetraploids, 180
Ammonites, parallel variation, 119
Anagenesis, 135
Anopheles maculipennis species-complex, 78
Anthropoidea, classification, 32
Aonidiella aurantii, resistant strains, 177
Apomixis, 99
Aristotle, 27, 28
Arkell and Moy-Thomas, 119
Armstrong, 173
Art, 69
Artenkreis, 69

Barnardius, 136, 166
Barriers, between species, 160
Bergmann, 151
Binomial (binominal, binary) system, 43
Biological races, 171
Biospecies, 121
Boarmia repandata, 158
Brehm, 78
Brigade, 31

CANADIAN Pondweed (*Elodea*), 98
Capercaillie and black grouse, hybrids, 95
Carabus cancellatus, 148
Carnivora, classification, 20
Cepaea nemoralis and *hortensis*, 76; overlap, 153
Chaffinch, double invasion, 139; and brambling as good species, 76
Characters, abstract nature, 147; adaptive, 21; in Carnivores, 22; definition, 147; diagnostic, 16, 25; geographical variation of significance, 65; non-adaptive, 26; of no importance, 147; specific and subspecific significance, 65
Chironomidae, 84
Chromosomes, 80, 143, 179
Citation, taxonomic, 70
Cladogenesis, 135
Class, 28, 31
Classification, artificial, 18; horizontal, 40; natural, 18; by resemblance and difference, 18; vertical, 40
Cleistogamy, 178
Clines, 149
Coefficient of selection, 146
Coexistence, artificial, 94; genetical, 93, 94
Cohort, 31
Convergence, 67
Creodonts, 23, 40
Crossability, 94; and gentes, 112
Crows, carrion and hoodie, hybrid zone, 95

DA Cunha, 156
Darwin, on geographical variation, 130; speciation, 169; species, 51, 182; taxonomists, 12
Daubentonia, 25
Diploid, 179
Divergence, of populations, 143
Dobzhansky, T., on agamospecies, 105; isolating mechanisms, 161; polymorphism, 157; speciation, 13
Drift, genetical, 144
Drosophila guaru, *guarani* and *subbadia*, 170; *polymorpha*, 156; *pseudoobscura* and *persimilis*, sibling species, 80; isolating mechanisms, 161
Ducula, characters, 66; on small islands, 167

ECHINODERMATA, 84
Ecological replacements, 88; compatibility, 164

Ecotypes, 170
Eiseniella tetraedra, parthenogenetic, 99; varieties, 101
Elton, C., 88
Empire, 28
Eurystomus orientalis, overlap of wintering subspecies, 91

FAMILY, 31
Fertility, 94
Fisher, R. A., 146
Fissipedia, 39
Ford, E. B., 166
Forma, 101
Formenkreis, 69
Fossils, and taxonomic difficulties, 38, 106

GAMETE, 179
Gause's principle, 164
Genes, 103; multiple effects, 148, 179
Genotype, 103
Gens, 107; forking of, 113; as evolutionary unit, 111
Genus, 28; horizontal and vertical, 119; polyphyletic, 120
Geospizinae, competition and beak size, 154
Glanville fritillary (*Melitaea cinxia*), 166
Gloger, 151
Grade, 31
Great Tit (*Parus major*), geographical variation, 55; ring species, 114, 141
Group, as general term, 31; polyphyletic, 120; delimitation and recognition of natural groups, 19, 25
Guillemot, cline in, 150
Gulls (*Larus*), ring species, 141

HABITAT-PREFERENCE, 91
Halcyon chloris, overlaps, 136
Haploid, 179
Hemicyon, 38
Heterozygote, 156
Hierarchy, 15; of rank-names, Linnean, 28; modern, 31
Homogeneity within groups, and ranks, 27
Homonyms, 46
Homozygote, 157
Huxley, J. S., 149
Hyaenarctos, 38
Hybridization, primary and secondary, 58, 88; and isolating mechanisms, 161; and specialization, 130; and species limits, 95; and gentes, 112
Hybrid zones, 95, 162

INDIVIDUALS, 35
Infraclass, 29, 31
Infraorder, 31
Insectivora, 23
Isolating mechanisms, 161; classified, 162

KEYS, 15
Kingdom, 28
Kleinschmidt, R., 69
Koopman, K., 161

LACK, D., 131, 154
Lakes and species-flocks, 174
Legion, 31
Linnaeus, and hierarchy, 28; on natural classification, 18
Lodge, 95
Lorenz, K., 156
Lumbriculus variegatus, vegetative reproduction in, 105

MAN, classification and affinities, 27, 29, 36
Mayr, E., on biological definition of the species, 86; characters, 148; *Eurystomus*, 91; fertility and crossability, 94, 112; guillemot, 150; hybrid zones, 162, 170; isolating mechanisms, 162; isolation by time, 87, 91; pseudoconspecific pairs, 164; polytypic species, 69; species-flocks, 174; superspecies, 70; sympatry and allopatry, 73; *Zosterops*, 92
Megatherium, 107
Melanism, industrial, 158
Miacids, 23, 38, 40
Micraster, evolutionary change in, 115
Monophagous, 171
Moore, J. A., 160
Moreau, R. E., 92
Morphology, guide to specific distinctiveness, 74; not always reliable, 75; subordinate to genetic criteria, 83

Morphospecies, 51, 121; characteristics, 52
Muller, C. H., 170
Multiple invasion, 138
Mutatio, 47, 101
Mutations, 103; random, 143

NAMES, arbitrary, 42; of families and subfamilies, 47; genera, 31, 43; subgenera, 45; species, 43; subspecies, 44; varieties, 47; trivial, 43, 44; standardized endings, 31; rank-names, 28; of particular groups, 27; inappropriate, 46; full citation of, 44
Neophema bourkii, 167
Neositta, 167
Newts (*Triturus*), overlap, 137
Nightingales (*Luscinia*), overlap, 137, 153
Nomenclature, binary or binomial, 43
Nuthatches (*Sitta*), convergence, 67; overlap, 154

OAKS, ecological selection in, 170
Oligophagous, 171, 172
Orchis traunsteineri and *fuchsii*, hybrids, 169
Order, 28, 31
Otocyon, 21, 24
Overlap of ranges, 90; in Darwin's Finches, 154; and competition, 152; and ecological specialization, 165; and speciation, 136

Pachycephala pectoralis, geographical variation, 63; hybridization, 138; overlaps, 64, 138
Paedogenesis, 85
Palaeospecies, 106, 121; defined, 107
Parentheses and intergeneric transfers, 44; and synonyms, 46
Parthenogenesis, 98; obligatory, 99; partial, 98, 104; recent, 104; and sympatric speciation, 177
Pentaploid, 179
Petersen, 174
Phylum, rank-name, 29, 31; evolutionary series, 107
Pieris (White butterflies), good species, 73; ecological replacement, 88
Pinnipedia, 38
Plasmodium strains, 171
Polymorphism, stable, 157
Polyphagous, 171
Polyploidy, 178; in asexual forms, 180
Polytelis, subgenera, 45; in Central Australia, 167
Populations, peripheral and ecological specialization, 165
Prairie chicken and sharptailed grouse, hybrids, 95
Primates, 25; classification, 29
Priority, Law of, 46
Proteles, 21
Ptarmigan and red grouse, ecological replacement, 89
Ptilinopus, variation 65; and coexistence, 154; on small islands, 167; multiple invasions, 139; complete overlaps, 138; zoogeography, 72

QUADRINOMIALS, 46

Rana pipiens, 160
Ranges, breeding, 91; contiguous, 93
Rank, 27; positional value, 37
Ranks, Linnean, 28; modern, 31; obligate, 29
Rassenkreis, 69
Ray, John, 18
Rensch, B., 69, 174
Replacement, geographical versus ecological, 88; and intrinsic factors, 92; and genetical sympatry, 93
Reproduction, sexual, and evolution, 103; vegetative, 98, 105
Ring species, 113, 141
Rodentia, 25
Rosella parrots (*Platycercus*), geographical variation, 62, 136; full citation exemplified, 70; superspecies, 70; overlap, 153
Rules, ecological, 151; of nomenclature, 41

SECTION, 31
Selection, evidence for, 149
Semispecies, 70
Series, geographical, 68; time, 40, 110
Shrews (*Sorex*, *Crocidura*), geographical variation, 55

Simpson, G. G., on classification of primates, 27, 29, 37; species and time, 111
Southern, H. N., 150
Sparrows (*Passer*), stabilized hybrid populations, 162
Speciation, geographical, 134; semi-sympatric, 169; sympatric, 169; by allopolyploidy, 180
Species, asexual, see agamospecies; biological, 73, 86, 93, 102; degrees of distinctiveness, 75, 82; description of new species, 49; 'good', 49, 73; gaps between, 49; and geographical variation, 68; interbreeding criterion, 83; hybridization, 95; and time, 110, *see* palaeospecies; monotypic, 52; morphological, *see* morphospecies; polytypic, 47; pragmatic definition, 51; sibling, 80; sibling and sympatric speciation, 175; recognizable on museum material, 82; specially created, 49; restricted to the biospecies, 124; interrelationships of different sorts, 120
Species-flocks, 172, 174
Species-groups, 31, 46, 70
Species-pairs, in Europe, 137
Stratigraphical *v.* biological classification, 108
Subclass, 30, 31
Subfamily, 29, 31
Subgenus, 31
Subkingdom, 30, 31
Suborder, 29, 31
Subphylum, 30, 31
Subspecies, 31, 69; and intergradation, 68
Sukatschew, 158
Superclass, 31
Superfamily, 29, 31
Superorder, 31
Superspecies, 70; as zoogeographical unit, 71
Swallowtail butterfly (*Papilio*), 166
Sympatry, 73; genetical, 91, 93
Synonyms, 46
Systema Naturae, 43

Tanysiptera, overlaps, 138
Tetraploid, 179
Thorpe, W. H., 172
Time-quantum, of the biospecies, 102
Tinbergen, N., 156
Transients, 111
Treecreepers (*Certhia*), 77, 137
Tribe, 31
Trichoglossus haematodus (coconut lory), geographical variation, 59; clines, 149, 150; rank, 87; overlaps, 155
Trinomials, 45
Triploid, 179
Trypanosoma, biological races, 171
Type of a species, 51

VARIATION, geographical, 53; apparently random, 147
Variety, 28, 31, 69; naming of, 47
Vaurie, C., 67

WAGNER, M., 132
Wallace, A. R., 152
Wren (*Troglodytes*), geographical variation, 53
Wright, S., 144, 145

ZOOGEOGRAPHY, and superspecies, 71
Zosterops, geographical replacement, 89, 93; multiple invasions, 139; overlaps, 153

#327